Electronic Assembly

Electronic Assembly

Jeremy Ryan

RESTON PUBLISHING CO., INC.
A Prentice-Hall Company
Reston, Virginia

Technical Drawings by Robert Mosher

Library of Congress Cataloging in Publication Data

Ryan, Jeremy
Electronic assembly.

Includes index.
1. Electronic apparatus and appliances—Design.
and construction. I. Title.
TK7870.R88 621.381 79–17953
ISBN 0–8359–1639–1
ISBN 0–8359–1638–3 pbk.

1 3 5 7 9 10 8 6 4 2

PRINTED IN THE UNITED STATES OF AMERICA

Contents

Preface

The reliability of any electrical or electronic equipment is dependent upon the knowledge and skills of the technicians that assemble the equipment. The finest engineering and the most costly equipment can be made worthless by a lack of ability or sloppy workmanship on the part of the assembler.

It is the purpose of *Electronic Assembly* to provide students of electronics with a sound background in all aspects of electronic assembly. This purpose is to be accomplished by treatment of the subject in a logical and clear, yet simplified manner. Exercises at the end of each chapter lead the student through graduated exercises to a thorough comprehension of the assembly process.

Techniques of assembly presented within this text are based upon the high standards of: NASA specifications, manufacturers such as Tektronics Corporation and Hewlett-Packard Company, and experiences of the author. The examples presented will aid the student in developing assembly skill and knowledge of materials. However, these skills can only be mastered by practice and in the final analysis, the reliability of electronic equipment is the responsibility of the individual assembler.

The author wishes to acknowledge those who preceded him by their preparation of other textbooks on this subject, and the many persons who aided him in the preparation of this work.

This text is respectfully dedicated as a teaching tool to the teachers of technical schools, community colleges, and high schools.

1: Safety

1.0 INTRODUCTION

An electronic assembler works with many complex electrical devices and electrical equipment that can be hazardous if used improperly. For this reason, if you want to work as an electronic assembler you must learn and follow the safety rules and procedures of your occupation. To help you learn to always use these rules and procedures we will consider them systematically in three parts: (1) general safety rules and procedures, (2) hand-tool safety rules and procedures, (3) soldering safety rules and procedures. This chapter explains the general rules of safety. The other two sets of rules are also presented here briefly and then in Chapter 4 (hand tools) and Chapter 7 (soldering) where their uses are treated by means of various examples. Remember to learn the Key Terms (KT's) also.

1.1 ELECTRIC SHOCK

Safe operating procedures are each person's responsibility and extend into many areas at home, on the highway, in the shop, and in the laboratory. In order to observe safety precautions in any particular place and especially with electricity, therefore, a person must know what the hazards are. Otherwise he/she easily becomes familiar as he/she works, and all too often, familiarity with the equipment leads to carelessness—with fatal results. So it is absolutely necessary to point out certain principles of electrical shock so that the need for, and the nature of, safety precautions may be thoroughly appreciated.

An important fact to remember is that *current rather than voltage determines shock intensity*. Of course, there must be voltage to cause current flow, but it is the way in which that current is resisted that determines the effect. The overall *resistance* properties of the human body as well as the point of resistance can vary from a few hundred *ohms* to many thousand ohms.

If the skin is dry, it presents a high contact resistance of several hundred thousand ohms to a voltage so that only a small current flows through the body. However, if the skin is damp or wet, the contact resistance can be less than 1000 ohms. In that case, even a low voltage can be fatal since a large current can pass through the body. In other words, a lower resistance allows a stronger current. The chart in Figure 1-1 shows the effects on the human body of different levels of current. The actual body contact resistance depends also upon the area of contact to a large degree.

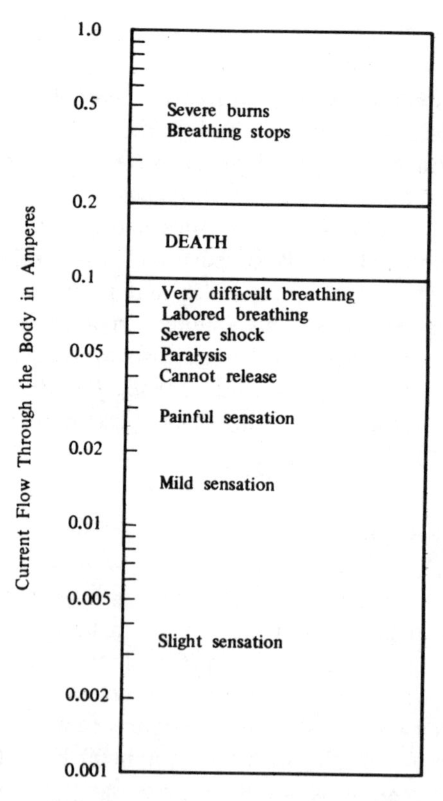

Figure 1-1 Degree of electric shock and accompanying body sensations

The best safety rule is to treat all *electrical circuits* as possible health *hazards*. Keep this in mind and always follow these safety procedures:

GENERAL SHOP RULES

1. Never work on electrical equipment alone.
2. Move slowly; make sure your feet are placed for good balance. Never *lunge* for falling tools.

Figure 1-2 Move slowly, make sure your feet are placed for good balance

3. Never work on live equipment when *fatigued* and always keep one hand in a pocket when working on live equipment.
4. Do not touch electrical equipment while standing on either a damp or metal floor.

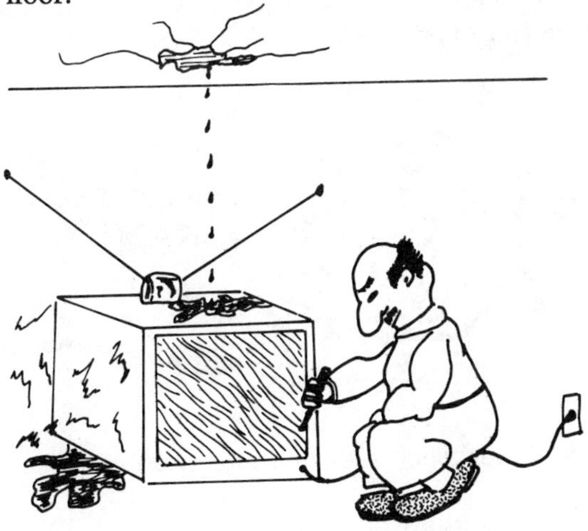

Figure 1-3 Do not touch electrical equipment while standing on a wet or damp floor

5. Do not handle electrical equipment while either you or the equipment is wet or damp.
6. Do not handle electrical equipment that is not *grounded.*
7. Clean up immediately any liquids spilled on workbench or floor.
8. Keep your work area neat and clean. Put waste material into proper receptacles.

Figure 1-4 Keep your work area clean

9. Do not use water on an electrical fire.
10. Never touch two pieces of equipment at the same time.
11. Never touch even one wire of an electrical line—it may be the one with voltage (the "hot" one).

Figure 1-5 Never touch even one wire of an electrical circuit

12. Do not take unnecessary risks.
13. Remove a victim from an electrical circuit as quickly as possible without endangering your safety. Any delay may prove fatal.

14. Start *artificial resuscitation* on an electrical shock victim at once and continue until the victim is revived or pronounced dead by a doctor.
15. Never carry bulky objects that obstruct vision.
16. Never attempt to repair or adjust any electrical equipment unless such a procedure is part of your duties.
17. Do not touch any part of any electrical equipment, fixtures, or wiring with wet hands.
18. Grasp the plug *not* the cord to remove the line-cord from an electrical outlet.
19. Wear short-sleeved clothing around drill presses and other revolving machines. Do not wear neckties, rings, bracelets, brooches, or anything that can become entangled in mechanical equipment. Keep long hair under a hat or tied back while you work.
20. Wear safety glasses when grinding, drilling or doing any other work where flying materials may injure the eyes.
21. If you are soldering on equipment above eye level, wear safety glasses also.

Figure 1-6 Always wear safety glasses or a face shield when soldering or using machine tools

22. "Good housekeeping" is an important factor in good safety at all times.
23. Finally, observe safety practices
    a. For self,
    b. For others,
    c. For equipment.

TOOL AND EQUIPMENT SAFETY

1. Before working on electrical equipment, turn off all power and ground all high-voltage points. Make sure that the power cannot be restored accidentally.
2. When you cut wires, keep the open side of the wire cutter away from your body and keep cut-off wire ends in your own area.
3. Keep a protective cover over knife blades.
4. Keep tools and materials from hanging over the edge of benches.
5. Place tools and materials so they cannot slide, roll, or fall.
6. Never work on rotating machinery with someone standing too close.
7. Never use a box, barrel, chair, or any other makeshift support as a substitute for a ladder.
8. Replace or repair splintered, broken, rough, or loose tool handles before use. Use only files that are equipped with handles.
9. Shield the dangerous parts of sharp-edged or pointed tools when you carry them from place to place.
10. Keep screwdriver blades sharp and even. (Do not use screwdrivers with broken or rounded points or broken handles.)
11. If a grinding wheel vibrates or wobbles, stop it at once and report the problem to your instructor or shop supervisor.
12. When you use a drill press, be sure to hold the work with a drill vise or some other clamping tool.

GENERAL SAFETY

1. It is wise to work with one hand at a time when making high voltage adjustments.
2. Never hold solder in your mouth when soldering. Your soldering iron can give you a shock through this low resistance path.
3. Avoid interfering with or startling anyone using a hot soldering iron or other equipment.
4. Lift soldering irons only by their handles.
5. Place your soldering iron so you don't have to reach across or around it when you resume work.
6. Keep your soldering iron in its proper holder when it is not in actual use.
7. When unsoldering a wire, make certain there is no tension or spring to the wire that could cause hot solder to be flipped into your face.
8. Never place rags, paper towels, etc. near a hot soldering iron.
9. Fumes of burning Teflon are poisonous. Solder Teflon only in a well-ventilated area.

Figure 1-7 Never startle anyone while he or she is working on electrical equipment

Figure 1-8 Place your soldering iron so that you do not have to reach across it

10. Do not breathe carbon tetrachloride fumes.
11. Never use flammable liquids, such as paint thinners, naphtha, and gasoline, in enclosed areas.
12. Wash your hands thoroughly after using any solvent.

KT's KEY TERMS TO REMEMBER

Artificial Resuscitation The method of reviving a person suffering from electrical shock by manipulating the diaphragm so as to force air into the lungs or by breathing air through the mouth (mouth-to-mouth resuscitation) into the lungs.

Current The movement of electrons through a wire or other conductor of energy; measured in amperes.

Electrical Circuit A number of conductors connected together in order to carry an electrical current; or the interconnection of devices in one or more closed paths to perform an electronic function, such as an amplifier that makes a music signal louder.

Fatigue The weakening of a material, such as metal, under repeated stress; in a human, tiredness.

Ground A metallic connection that has the same voltage level as earth and can be an actual connection to it. The metal case of a piece of equipment is usually grounded by a neutral electrical lead.

Hazards Operations or equipment that can hurt or destroy if they are used incorrectly or if they malfunction.

Lunge To move forward too suddenly, and usually so as to be off balance.

Resistance The property of a material or the human body that limits the passage of electrical current.

Voltage The electrical pressure or force that causes current to flow through any conductor of electricity.

SAFETY TEST #1

All students must complete this test. Correct answers are necessary before you continue studying.
Circle either "True" or "False" for each statement.

	(Circle the correct term)	
1. You should never work on energized electrical equipment alone.	T	F
2. You may be severely shocked (complete an electrical circuit through your body) if you stand on a damp or metal floor while working on an electrical device.	T	F
3. It is safe to work on electrical equipment if it, or my body, is damp.	T	F

(Circle the correct term)

4. For safety, all electrical equipment should be grounded.	T	F
5. Water should be used on an electrical fire.	T	F
6. It is safe to throw any waste material into any waste container.	T	F
7. To avoid completing a circuit, *never* touch two pieces of electrical equipment at the same time.	T	F
8. It is safe to touch only one wire at a time of an electrical circuit.	T	F
9. An electrical shock victim should be pulled away from the circuit without touching their body.	T	F
10. Artificial resuscitation should be started at once for an electrical shock victim.	T	F
11. Repairs or adjustments to equipment should be done only when it is an assigned duty.	T	F
12. Electrical cords should be unplugged by pulling on the wire.	T	F
13. Loose clothing, long hair, and jewelry should be worn in the shop if they are fashionable.	T	F
14. Eye protection masks (goggles) must be worn when doing anything that may cause particles to hit the eyes.	T	F
15. Eye protection is optional when soldering above eye level.	T	F
16. Good housekeeping is part of good safety practices.	T	F

SAFETY TEST #2

Circle either "True" or "False" for each of the following statements.

1. It is a good idea to hold solder in your mouth when soldering.	T	F
2. You should not startle anyone in the shop area.	T	F
3. Soldering irons should be picked up by their handles even if you think they are cold.	T	F
4. The soldering iron tip should be positioned so arms or other parts of the body do not come close to it when you reach to use it.	T	F
5. A soldering iron can be safely left anywhere on a bench.	T	F

6.	Wires being unsoldered can cause solder to spray and cause burns.	T	F
7.	You should always connect the ground wire before energizing a circuit.	T	F
8.	You should make high-voltage adjustments by placing both hands on the equipment.	T	F
9.	Burning insulation such as Teflon produces poisonous gases.	T	F
10.	Enclosed areas are best for using flammable liquids.	T	F
11.	You don't have to wash your hands if you have been working with solvents.	T	F

SAFETY TEST #3

Good practice

Bad practice

Good practice

Bad practice

Good practice

Bad practice

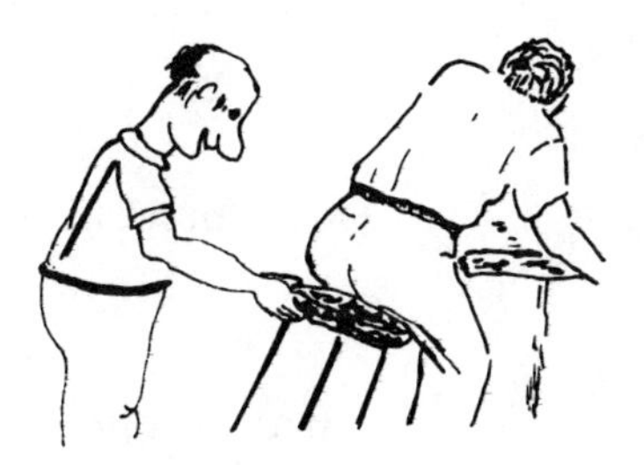

Good practice

Bad practice

Good practice

Bad practice

Good practice

Bad practice

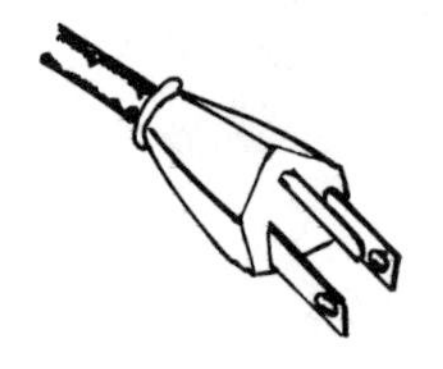

Good practice

Bad practice

Good practice

Bad practice

Good practice

Bad practice

2: Parts Identification

2.0 INTRODUCTION

Electrical and electronic components are those parts connected to form an *assembly*. Working together, they produce a desired *output*. In the following pages we present a list, pictures, and symbols of the most often used components in addition to the technique for identifying components by color code. As a basis upon which to learn electronics, you must learn to identify these components by appearance, circuit symbol, and description. To better understand the material in this chapter you may need to refer to the glossary of terms as you read. Italicized words are those found in the glossary. Remember to learn the KT's at the end of this chapter.

2.1 COMPONENT SYMBOLS, APPEARANCE, AND DESCRIPTION

An antenna (Figure 2-1) is an exposed conductor whose function is to intercept or radiate *electromagnetic* waves.

Figure 2-1 Antenna

Batteries (Figure 2-2) are made in many shapes and sizes using several combinations of *conductors* and *electrolytes.* Each battery consists of two or more cells connected to produce direct current voltage. A cell is a single unit capable of generating an electric current.

Figure 2-2 Battery containing cells

Capacitors (Figure 2-3) are characterized by the ability to store electric charges by means of an *electric field.* This property is called *capacitance.* Capacitors are generally grouped into three categories: paper or dry, electrolytic or wet, and air or *vacuum.*

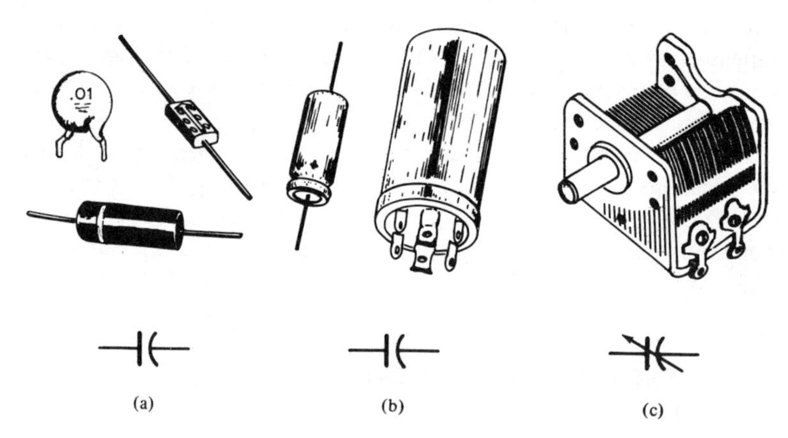

Figure 2-3 Capacitors: **(a)** fixed, **(b)** electrolytic, **(c)** variable

An inductor or coil (Figure 2-4) is a series of loops or turns of wire that acts by means of a *magnetic field* to oppose any change in the existing current in a circuit. This opposition is called *inductance.* The inductor in Figure 2-4 a has an air core and is fixed. The inductor in Figure 2-4 b has a variable metal core.

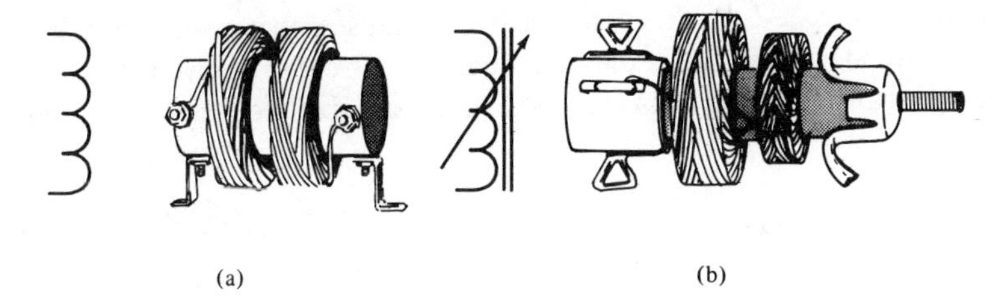

Figure 2-4 Inductors (coils): **(a)** fixed, **(b)** variable

Figure 2-5 shows symbols that represent various conductors: (a) shows the symbol for two conductors that are crossed but not connected, (b) connected conductors, and (c) a *shielded conductor*.

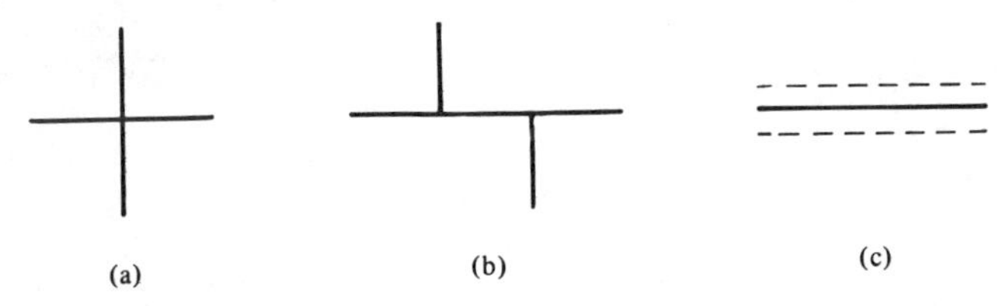

Figure 2-5 Conductor symbols: **(a)** not connected, **(b)** connected, **(c)** shielded

The diode or rectifier (Figure 2-6) is a device that conducts *electrons* more easily in one direction than the other. It can serve as a one way *switch* which closes to permit current flow in only one direction.

Figure 2-6 Diode (rectifier)

A fuse (Figure 2-7) is a device sensitive to the amount of current being forced through it. When the amount of current is too great, an internal element breaks causing the circuit to open. This stops current flow just like a switch. An automatic fuse, called a circuit breaker, is shown in Fig. 2-7 b. Excessive current causes the circuit breaker to "trip" open. After the excessive load has been removed the circuit breaker can be reset. This type of "fuse" can be used again and again.

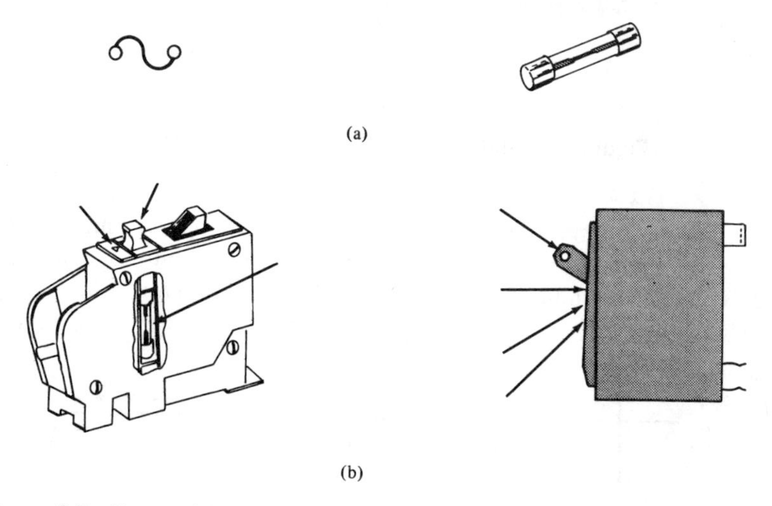

Figure 2-7 Fuses: **(a)** fuse & symbol, **(b)** electromatic fuse called a circuit breaker

Figure 2-8 shows symbols used to represent common grounds to the earth or the chassis. The chassis is the framework on which the components are mounted.

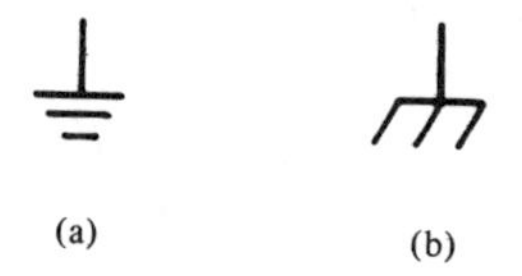

Figure 2-8 Ground symbols: **(a)** chassis, **(b)** earth

An *integrated circuit* (IC) (Figure 2-9) is a solid-state device that contains in one unit several components such as resistors, capacitors, and transistors. These components are produced on a small piece of *semiconductor* material called a chip. An IC performs a specific electrical function such as an amplifier.

Figure 2-10(a) shows a phone jack, (b) a plug, and (c) a receptacle for house current. The receptacle is similar to the phone jack but much larger. They all provide electrical connections that can easily be connected or disconnected by removing the pronged end.

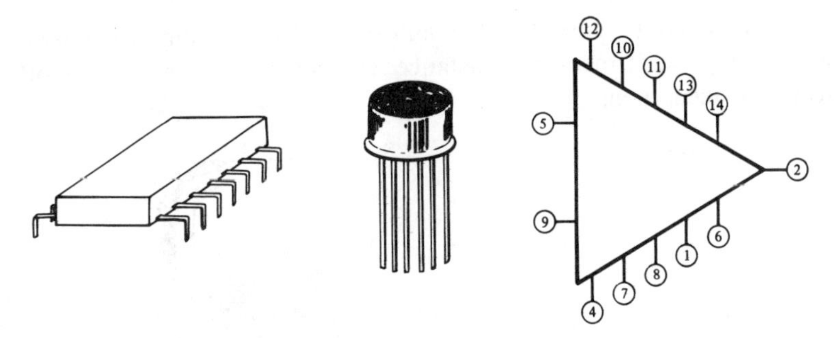

Figure 2-9 Integrated circuits

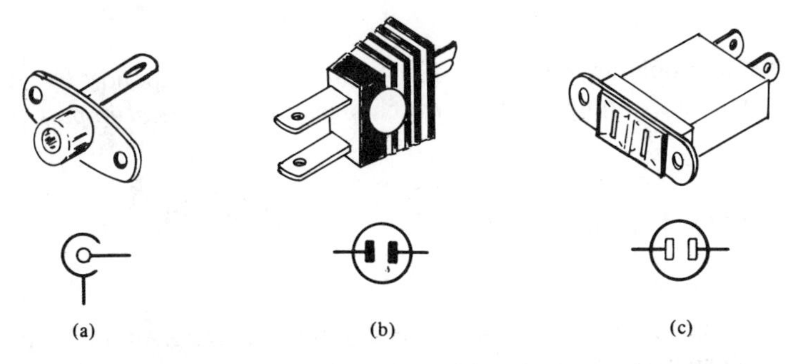

Figure 2-10 **(a)** phone jack, **(b)** plug, and **(c)** receptacle

Figure 2-11 shows various types of bulbs: (a) an illumination or pilot bulb, (d) neon bulb, and (c) a light-emitting diode. The element in an illumination bulb is a continuous circuit that ceases to function when it breaks. A neon bulb contains two elements and is filled with neon gas. The gas permits conduction from one element to the other. The light-emitting diode is used as a small indicator light. You have seen these on such instruments as calculators.

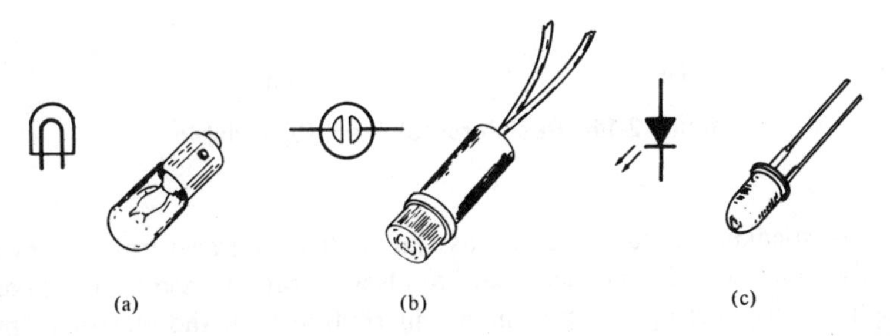

Figure 2-11 Bulbs: **(a)** illumination (pilot), **(b)** neon, **(c)** light-emitting diode (LED)

The meter (Figure 2-12) is an electrical measuring instrument. It can be designed to measure resistance, current, voltage, power, or other electrical phenomena.

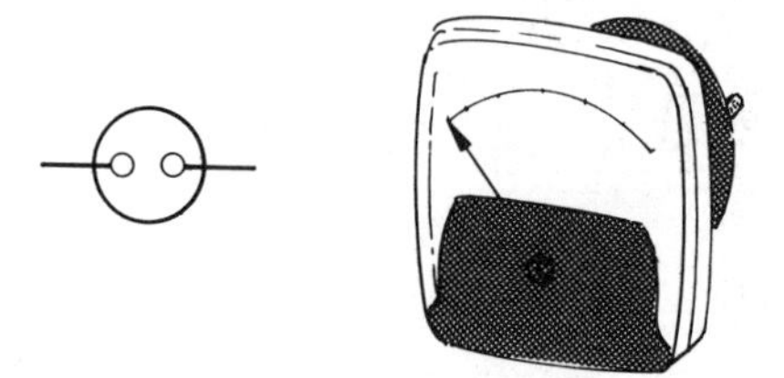

Figure 2-12 Meter

A microphone (Figure 2-13) is a unit that responds to sound waves by transforming *air pressure* changes into *electrical impulses.*

Figure 2-13 Microphone

A resistor (Figure 2-14) is a component designed to limit the flow of current in a circuit. This property is called resistance. Figure 2-14 shows the two types of resistors: (a) fixed, and (b) variable or potentiometer. Fixed resistors provide an unadjustable amount of resistance whereas variable resistors are manually adjustable.

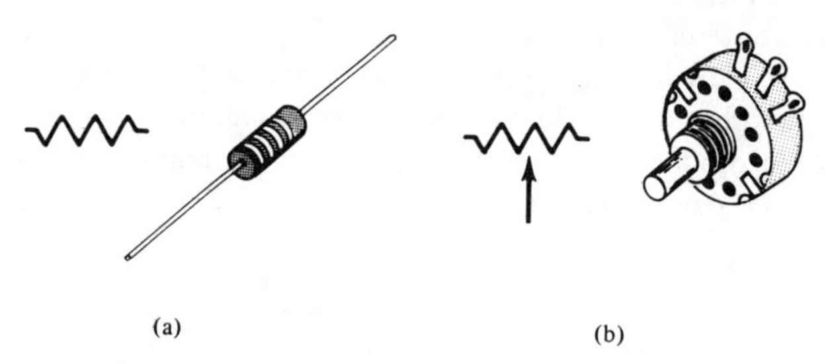

Figure 2-14 Resistors: **(a)** fixed, **(b)** variable

A speaker (Figure 2-15) has the ability to transform electrical impulses into sound. The electrical impulses vibrate a cone in a speaker which in turn sets up air movements in rhythm with the electrical impulses. These air movements cause us to hear what's being fed into the speaker.

Figure 2-15 Speaker

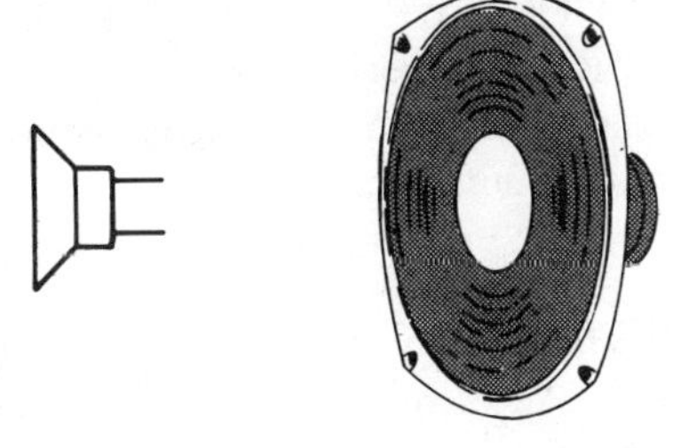

A switch is a device that permits the connection of a circuit on a temporary basis by completing or breaking the path of the electric current. There are many varieties of switches to serve many purposes. Figure 2-16 (a) shows a single-pole, single-throw switch (SPST), (b) shows a circuit symbol of a double-pole, double-throw model (DPDT), (d) shows a wafer switch which has the advantage of providing several contact points so that several circuits may be completed or broken. Wafer switches are generally made to handle smaller electrical demands than the toggle switches. Toggle switches (Fig. 2-16c) are activated by an up and down action of the switch lever. Wafer switches, on the other hand, are operated by a rotating action.

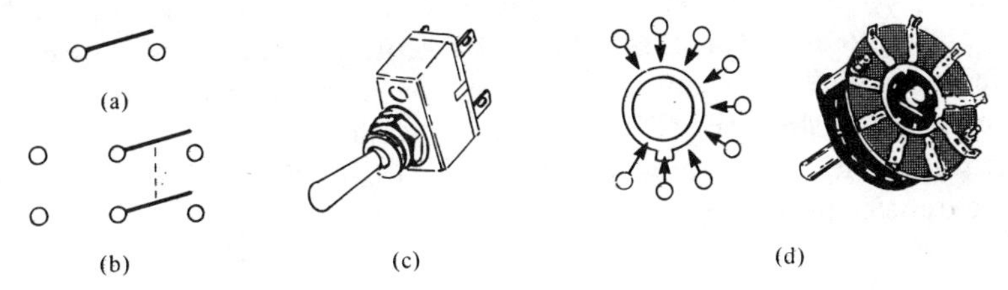

Figure 2-16 Switches: **(a)** SPST, **(b)** DPDT, **(c)** toggle switch, **(d)** wafer switch

A terminal is a device used as a point of connection for the wires and components in electrical circuits. Figure 2-17 shows three types of terminals: (a) bifurcated, (b) turret, and (c) hook.

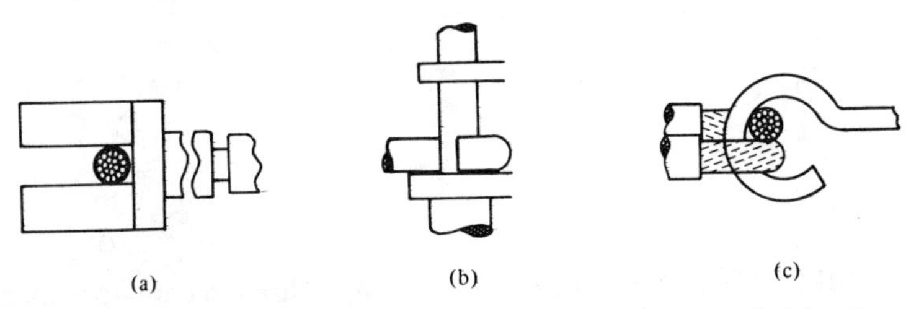

Figure 2-17 Terminals: **(a)** bifurcated, **(b)** turret, **(c)** hook

A transistor (Figure 2-18) is a solid-state device that performs the functions of a vacuum tube in addition to many functions the vacuum tube cannot perform. Furthermore, it is economical to operate because of its low power consumption.

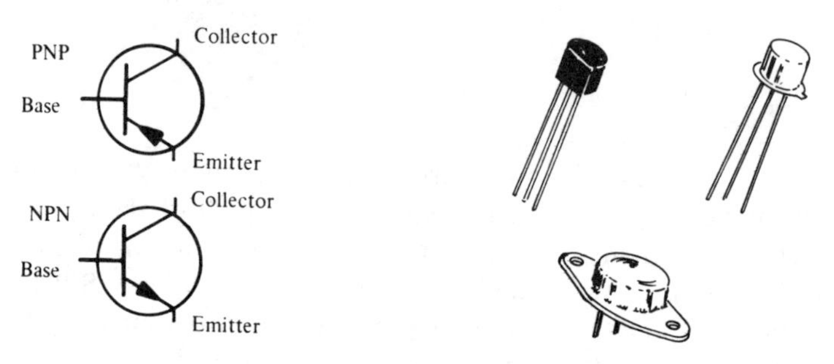

Figure 2-18 Transistors

When two coils are wound in close *proximity* to one another, they become a transformer (Figure 2-19). A transformer has the ability to transfer the effects of a current change from one coil to another. The primary coil is connected to the source of energy; the secondary coil is connected to a load such as a light or motor. The core of a transformer affects its application and its function; for example, power circuits use iron-core transformers (Figure 2-19 b), while radio and television circuits generally use air-core transformers (Figure 2-19 a), or adjustable core transformers (Figure 2-19 c).

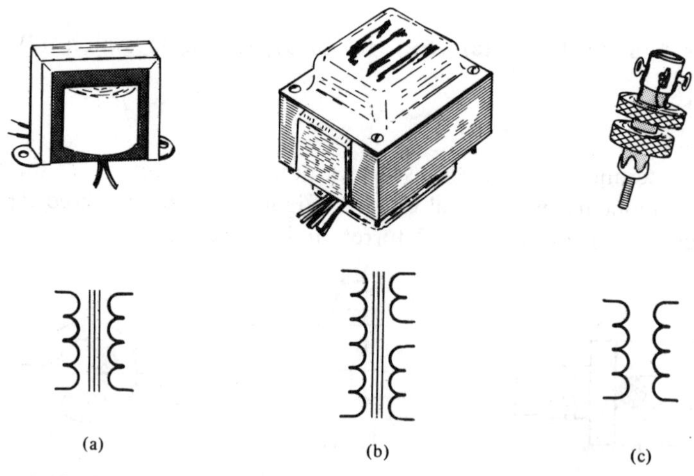

Figure 2-19 Transformers: **(a)** air core—primary, **(b)** iron core—secondary, **(c)** adjustable core

A vacuum tube (Figure 2-20) is designed to enable conductors to perform specific functions in a *vacuum*. The number and type of conductors or elements within a vacuum tube determine its function; for example, gas tubes are designed to function with a gas content within the glass envelope.

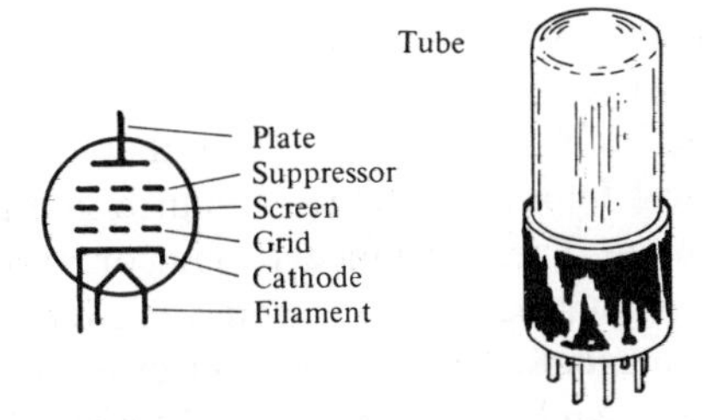

Figure 2-20 Vacuum tube

A binding post such as shown in Figure 2-21 is used to make temporary connections of leads or wires to an electric unit.

Figure 2-21 Binding post

A light-emitting diode (Figure 2-22) is used as a signal indicator. This device uses much less energy than an incandescent lamp and operates on much less voltage than the neon bulb.

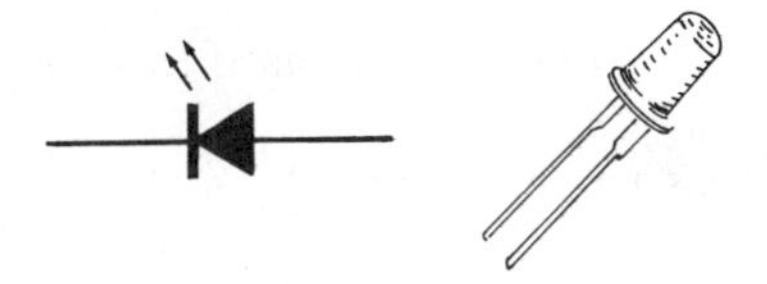

Figure 2-22 Light-emitting diode (LED)

The integrated circuit shown in Figure 2-23 may contain many transistors and diodes. Some integrated circuits contain capacitor and resistor elements. These devices allow manufacturers to greatly reduce the physical size of electronics equipment.

Figure 2-23 Integrated circuit (IC)

KT's KEY TERMS TO REMEMBER

Air Pressure A force (energy) caused by the movement of air.

Assembly A group of components comprising a complete unit designed to perform a certain electrical function.

Capacitance The property of a capacitor to store electrical energy.

Conductor A material, usually a wire, through which current can move easily.

Electric Field An invisible force that surrounds a charged body.

Electrical Impulse A sudden change in a level of current or voltage.

Electrolyte The substance (liquid or paste) that forms the conducting medium between electrodes in a battery.

Electromagnetic Having both electric and magnetic properties. Television and radio reception is possible because of electromagnetic waves.

Electrons The parts of atoms that orbit the nucleus. Electrons carry an electric charge and thus serve as transporters of electricity.

Inductance The property of an inductor that enables it to store energy and causes it to oppose any change of current.

Magnetic Field An invisible field about a magnet or wire through which current is passing.

Output In an electrical circuit, the current, voltage, power, or driving force used to drive a speaker or any other device (transducer).

Proximity The state of being close to or close by.

Semiconductor A solid or liquid material, with a resistivity between that of metals and insulators, in which increasing temperature affects change concentration. Semiconductors are used to make diodes, transistors, and IC's.

Shielded Conductor A conductor enclosed in a metal case, used to keep out stray signals.

Vacuum An enclosed space from which as much as possible of the air and gases has been removed.

PROJECT: IDENTIFICATION OF PARTS

1. ________ Battery
2. ________ Meter
3. ________ Speaker
4. ________ Ground (chassis)
5. ________ Phone jack (large)
6. ________ Inductor air core (coil)
7. ________ Binding post
8. ________ Neon bulb
9. ________ Receptacle
10. ________ Fuse
11. ________ Resistor (fixed)
12. ________ Capacitor (electrolytic)
13. ________ Diode (rectifier)
14. ________ Potentiometer
15. ________ Power transformer
16. ________ Integrated circuit
17. ________ Transistor
18. ________ Switch (wafer)
19. ________ Conductors, crossing
20. ________ Microphone
21. ________ Light-emitting diode
22. ________ Vacuum tube
23. ________ Antenna
24. ________ Transformer (air core)
25. ________ Capacitor (dry)
26. ________ Switch (toggle)
27. ________ Capacitor (variable)
28. ________ Connectors,
29. ________ Illumination bulb
30. ________ Switch (DPDT)

Materials

1. Box of numbered electronic parts to identify
2. Electronics dictionary

Procedure

Identify each of the parts furnished to you by your instructor: by name, number, circuit symbol, and with an electrical function description such as in the following example:

Example: Diode—A diode is an electrical device used to convert ac current or voltage to dc current or voltage. The diode passes current more easily in one direction than in the other and thus may be used as a switch.

3: Color Codes of Electronic Components

3.0 INTRODUCTION

In Chapter 2 you learned to identify many electronic components by name and circuit symbol. In this chapter you will learn that the value of some components is often indicated on the component by a *code* of colors. These basic colors and the digit each color represents are given in Figure 3-1. Remember to learn the KT's at the end of the chapter.

3.1 RESISTOR CODING

The unit of resistance is the ohm, symbolized by the Greek letter omega (Ω). For example, a resistor might have a value of 470 ohm or 470Ω.

Three-Band Resistor Color Coding

Resistors not marked with a numerical value will have three, four or five color bands. Figure 3-2 shows a resistor with three color bands. The color band nearest the end always represents the first *significant figure;* and the third, the multiplier (number of zeros to be added). For example, if the resistor in Figure 3-2 were coded brown, green, and red, the value would be

Brown = 1
Green = 5
Red = 2 zeros (or 00)
1500 ohm

Color	Significant Figure	Decimal Multiplier	Resistance Tolerance in Percent
Black	0	1	±0
Brown	1	10	±1
Red	2	100	±2
Orange	3	1,000	±3
Yellow	4	10,000	±4
Green	5	100,000	±5
Blue	6	1,000,000	±6
Violet	7	10,000,000	±7
Gray	8	100,000,000	±8
White	9	1,000,000,000	±9
Gold	-	0.1	±5
Silver	-	0.01	±10
No color	-		±20

Figure 3-1 Basic color code

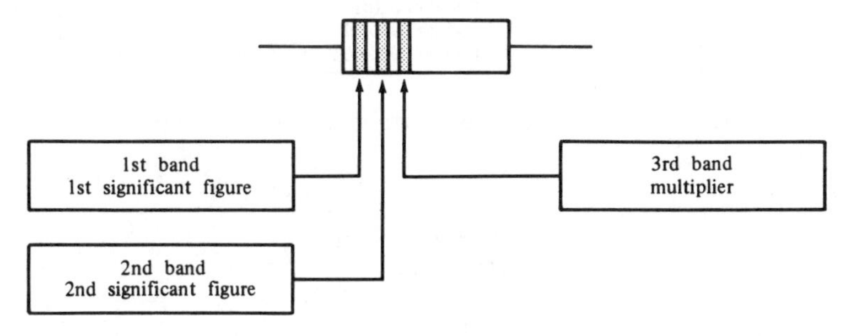

Figure 3-2 Three-band code for ± 20 percent resistors only

We would change this to read 1.5 Kohm, or 1.5 KΩ, by moving over the decimal and using the proper *prefix* as shown in Figure 3-3. The *tolerance* of a three-band resistor is always ±20 percent. That is, a resistor that reads brown, black, brown would be 100Ω ±20 percent; therefore, it would have a value somewhere between 80 and 120Ω.

Four-Band Resistor Color Coding

The four-band code arrangement shown in Figure 3-4 is used for resistors with a tolerance of ±1 to ±10 percent. The fourth band may be gold for ±5 percent, silver for ±10 percent, or any one of the colors for a tolerance between ±1 and ±10 percent. For example, a yellow, violet, red, gold resistor reads 4700 ±5 percent, or 4.7 KΩ. The value of this resistor is between 4.465 and 4.935 KΩ; however, you would ask for a

Power of ten	Number	Prefix	Symbol
10^{12}	1,000,000,000,000	tera	T
10^{11}	100,000,000,000		
10^{10}	10,000,000,000		
10^{9}	1,000,000,000	giga	G
10^{8}	100,000,000		
10^{7}	10,000,000		
10^{6}	1,000,000	mega	M
10^{5}	100,000		
10^{4}	10,000		
10^{3}	1,000	kilo	k
10^{2}	100	hecto	h
10^{1}	10		
10^{0}	1	deka	dk
10^{-1}	.1	deci	d
10^{-2}	.01	centi	c
10^{-3}	.001	milli	m
10^{-4}	.000 1		
10^{-5}	.000 01		
10^{-6}	.000 001	micro	μ
10^{-7}	.000 000 1		
10^{-8}	.000 000 01		
10^{-9}	.000 000 001	nano	n
10^{-10}	.000 000 000 1		
10^{-11}	.000 000 000 01		
10^{-12}	.000 000 000 001	pico	p

Figure 3-3 Prefixes are shortcuts for writing numbers

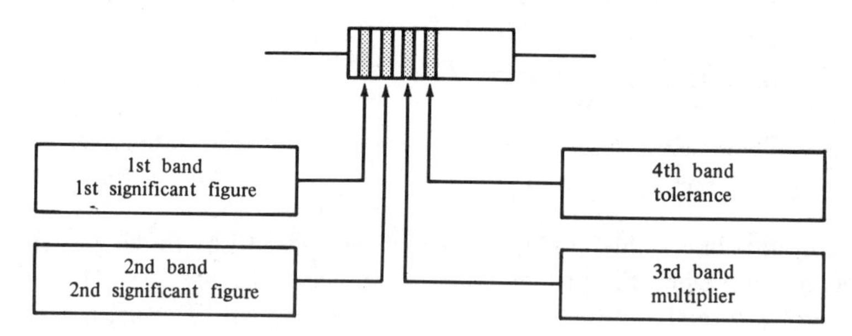

Figure 3-4 Four-band code for resistors with $\pm$ 1 to $\pm$ 10 percent tolerance

4.7 KΩ ±5 percent resistor at the supply room and a *schematic* would call for a 4.7 KΩ resistor.

Five-Band Resistor Color Coding

There are two five-band methods of coding resistors presently in use. The first is for resistors of a low tolerance ±1 or ±10 percent

(Figure 3-5). The first band denotes the first significant figure, the second the second significant figure, the third the third significant figure, the fourth the multiplier, and the fifth the tolerance. For example, a resistor with color bands red, green, orange, red, brown would read 25,300Ω ±1 percent. *Note:* This method allows a reading to the third place.

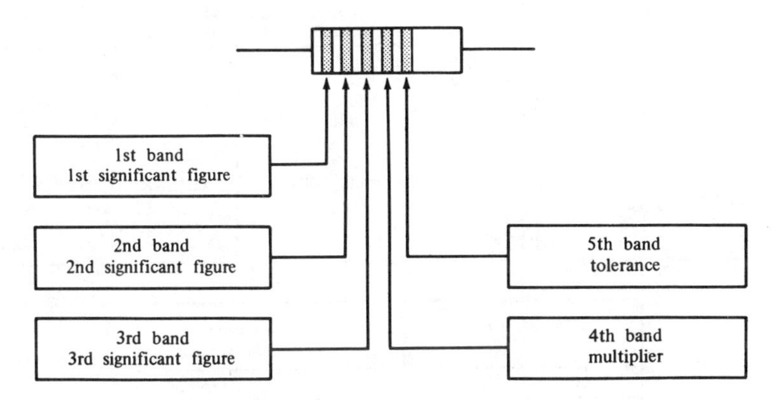

Figure 3-5 Five-band code for resistors with ± 1 to ± 10 percent tolerance

The other five-band type of resistor coding shown in Figure 3-6, is the same as the four-band coding but with a fifth band added for *reliability,* that is, the number of failures in 1000 hours. This fifth band is useful to the engineer in designing circuits. For example, an orange, orange, red, gold, brown code would indicate that the resistance value is 3300Ω ±5 percent. With a possibility of a one in one hundred failure rate in 1000 hours.

Noncolor Resistor Coding

The value of *precision* resistors, those with 0.5 or 1 percent tolerance, is often printed on the resistor, sometimes in a coded form. Figure 3-7(a) shows a 1 percent resistor with its value printed on the side. Figure 3-7(b) shows another 1 percent resistor with the value printed in a code. The F3462 is read 34,600Ω. The first three numbers are the first, second, and third digits; the fourth number is the multiplier.

The physical size and shape of a resistor determines its power rating. Ask your instructor to compare resistors of different power (wattage) ratings.

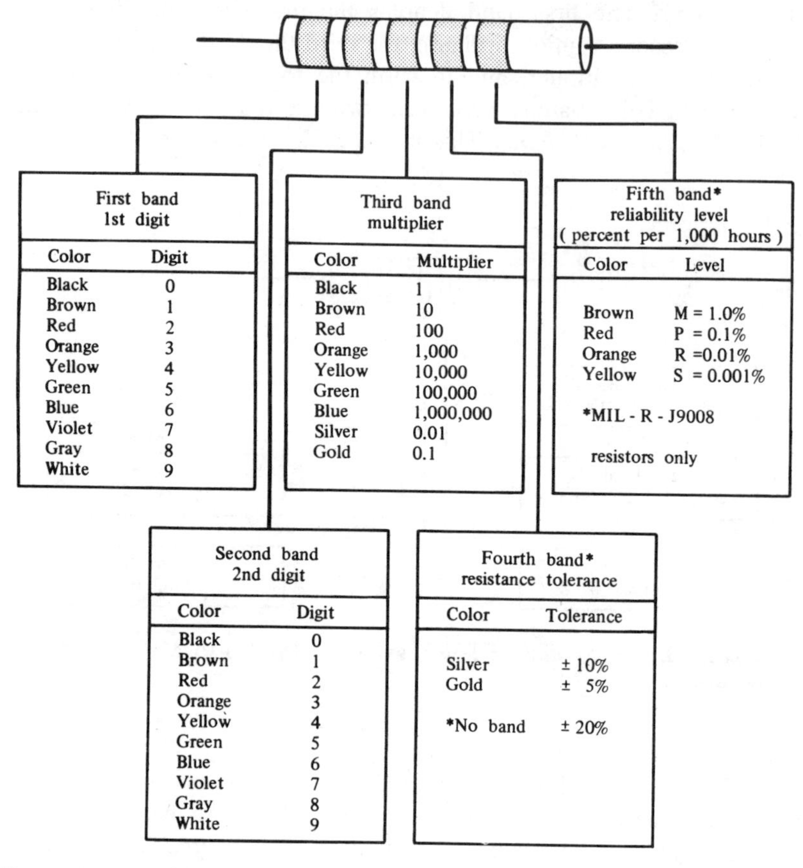

Figure 3-6 Five-band code for resistors with ± 1 to ± 10 percent tolerance (fifth band, reliability level)

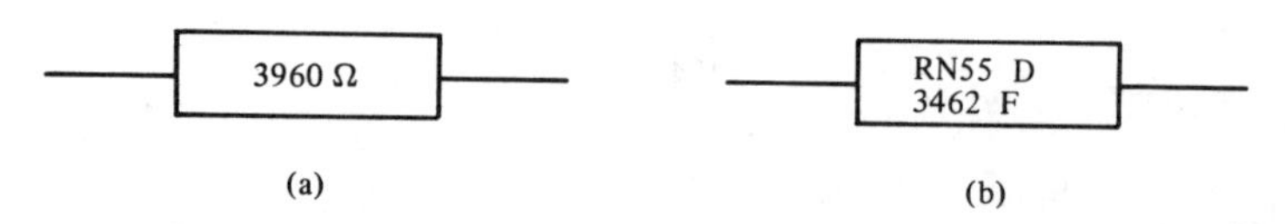

Figure 3-7 One percent resistor coding: **(a)** value written on side, **(b)** value in a numerical code

3.2 CAPACITOR COLOR CODING

There are several methods in use for coding capacitors. We will cover those methods you will most often encounter.

Coding Molded Tubular Capacitors

The color code for molded tubular capacitors is shown in Figure 3-8(a). In (b) the first two bands indicate the first and second significant figures, respectively, the third band the multiplier, the fourth band the tolerance. The fifth and sixth bands indicate significant figure ratings of a capacitor's maximum voltage with the fifth band denoting voltage in the hundreds and the sixth voltage in the thousands. For values less than 1000 V the sixth band is omitted. For example, a brown fifth band with a green sixth band indicates 1500 volts; a green fifth band by itself is 500 V.

Color	Significant Figure	Decimal Multiplier	Tolerance Percent (%)
Black	0	1	20
Brown	1	10	-
Red	2	100	11
Orange	3	1,000	30
Yellow	4	10,000	40
Green	5	10^5	5
Blue	6	10^6	-
Violet	7	-	-
Gray	8	-	-
White	9	-	10

Note: Voltage rating is identified by a single-digit number for ratings up to 900V, a two-digit number above 900V. Two zeros follow the voltage figure.

(a)

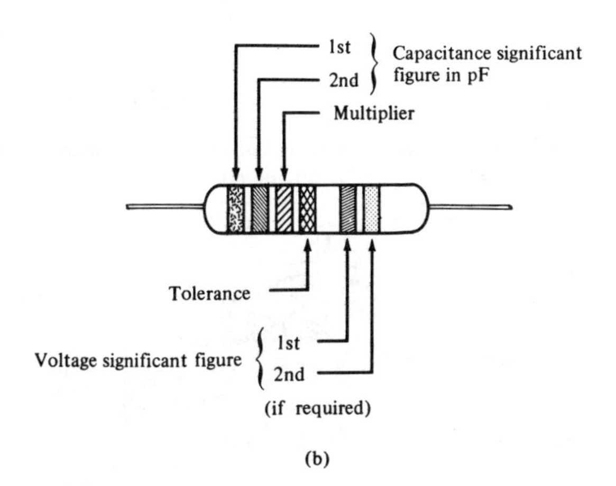

(b)

Figure 3-8 Color coding of molded tubular capacitors: **(a)** color code, **(b)** details of capacitor *(courtesy of Sprague Electric Company)*

Mica Capacitor Color Coding

There are three methods for coding mica capacitors. The *JAN* code in Figure 3-9(a) uses a black dot to indicate the beginning point of the code. The colors of the next two dots indicate the significant figures, the color of the third dot (in a clockwise direction) is the multiplier, the fifth the tolerance, and the sixth the characteristic of the capacitor. For example, using the color code in Figure 3-9(d), a capacitor with colors black, yellow, violet, red, green and brown equals *4700 pF* (picofarads), 5 percent tolerance, and 5000 volts rating, B characteristic.

The *RETMA* code in Figure 3-9(b) differs by beginning with a white dot. The colors of the second and third dots indicate the significant figures, the fourth the multiplier, the fifth dot the tolerance, and the sixth color the characteristic.

The code of the bottom silver capacitor in Figure 3-9(c) uses the colors of the first three dots to denote the significant figures, the fourth the multiplier, the fifth the tolerance, and the sixth the characteristic.

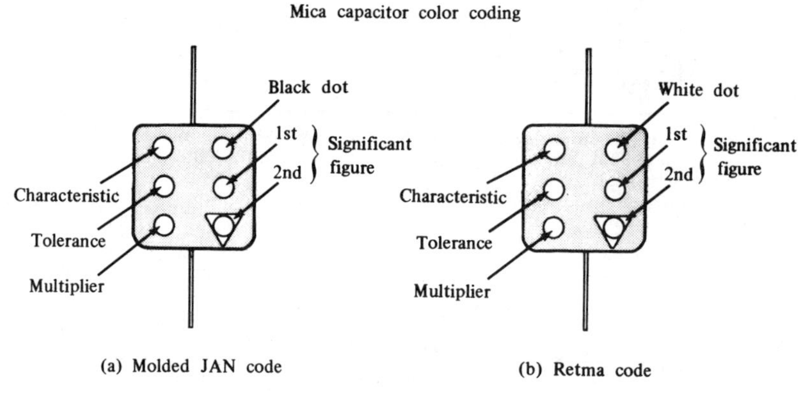

Note : Capacitance in pF. If both rows of dots are not on one face, rotate capacitor about the axis of its leads and read second row on side or rear.

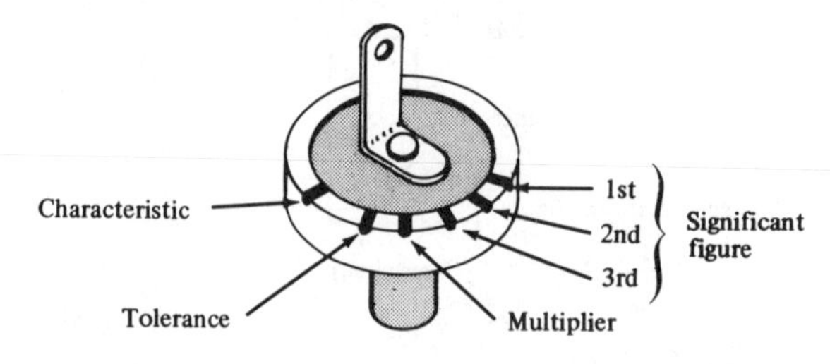

Note : Capacitance in pF.

(c) Button silver code

Figure 3-9 Mica capacitor color coding

Color	Significant Figure	Multiplier	Tolerance ± %	Characteristic
Black	0	1	20	A
Brown	1	10	-	B
Red	2	10^2	2	C
Orange	3	10^3	3 (RETMA)	D
Yellow	4	10^4	-	E
Green	5	-	5 (RETMA)	F (JAN)
Blue	6	-	-	G (JAN)
Violet	7	-	-	-
Gray	8	-	-	I (RETMA)
White	9	-	-	J (RETMA)
Gold	-	.1	5 (JAN)	-
Silver	-	.01	10	-
None	-	-	20 (RETMA)	-

(d) Color code

Figure 3-9 (continued)

Note: The numbering of all electronic components begins at a keyed point and runs clockwise, looking at the bottom of the component, or counterwise, looking at the top of the component.

Ceramic Capacitor Color Coding

As you can see from Figure 3-10, there are many ways of coding *ceramic* capacitors. To try to remember all of them would be very difficult. We suggest you refer to an electronic reference book when using ceramic capacitors.

Paper Capacitor Color Coding

Paper capacitors are coded in the three methods shown in Figure 3-11. Study these methods carefully and compare them to Figures 3-8, 3-9, and 3-10. The coding in Figure 3-11(b) is the Radio-engineers code and the code in Figure 3-11(c) is a joint Army-Navy code.

3.3 DIODE COLOR CODING

Diode types are indicated by three color bands read from the *cathode* end (Figure 3-12). Not all diodes will be color coded. Diodes that are not color coded usually have their type written on the body. Decoded, each type begins with 1N followed by a series of numbers. For example, if a diode in Figure 3-12 were coded red, violet, green, the type would be 1N275. The characteristics of the diode could then be found in the manufacturer's data book under that number.

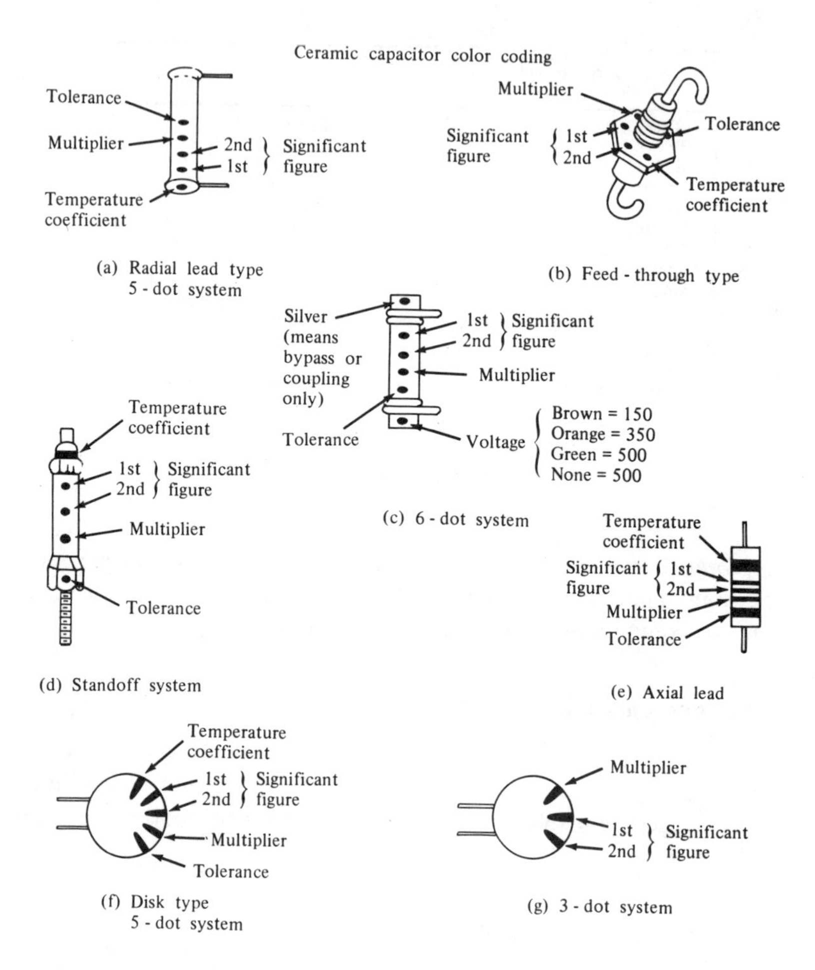

Color	Significant Figure	Multiplier	Tolerance A	Tolerance B	Temperature Coefficient
Black	0	1	2	20	0
Brown	1	10	.1	1	−30
Red	2	10^2	-	2	−60
Orange	3	10^3	-	2.5	−150
Yellow	4	10^4	-	-	−220
Green	5	-	5	5	−330
Blue	6	-	-	-	−470
Violet	7	-	-	-	−750
Gray	8	.01	.25	-	+30
White	9	.1	1	-	+120 to −750 (RETMA) +500 to −330 (JAN)
Gold	-	-	-	-	+100
Silver	-	-	-	-	Bypass or coupling

(h) Color code

Figure 3-10 Ceramic capacitor color coding

Paper capacitor color coding

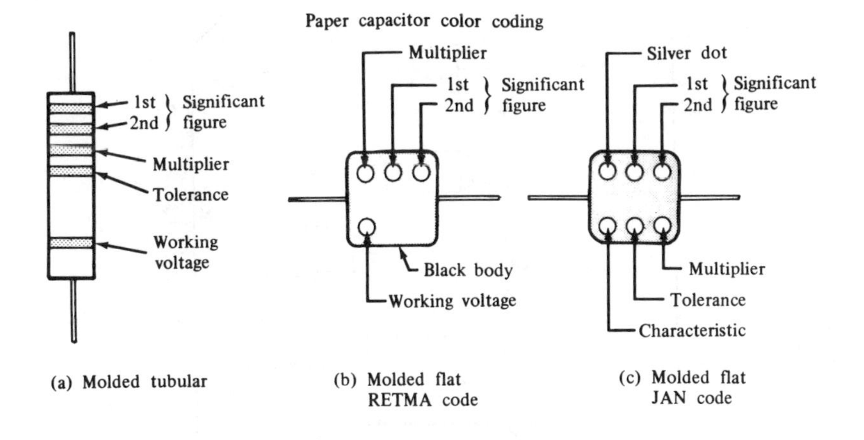

(a) Molded tubular

(b) Molded flat RETMA code

(c) Molded flat JAN code

Color	Significant Figure	Multiplier	Tolerance ± %
Black	0	1	20
Brown	1	10	-
Red	2	10^2	-
Orange	3	10^3	30
Yellow	4	10^4	40
Green	5	10^5	5
Blue	6	10^6	-
Violet	7	-	-
Gray	8	-	-
White	9	-	10
Gold	-	1	-

Note: Capacitance in picofarads. Working voltage coded in terms of hundreds of volts. Ratings over 900V expressed in two-band voltage code.

(d) Color code

Color	Significant Figure	Multiplier	Tolerance ± %	Characteristic
Black	0	1	20	A
Brown	1	10	-	B
Red	2	10^2	-	C
Orange	3	10^3	30	D
Yellow	4	10^4	40	E
Green	5	10^5	5	F
Blue	6	10^6	-	G
Violet	7	-	-	-
Gray	8	-	-	-
White	9	-	10	-
Gold	-	1	5	-

Note: Capacitance in picofarads. Working voltage coded in terms of hundreds of volts. Voltage ratings over 900V expressed in two-dot voltage code.

(e) Color code

Figure 3-11 Paper capacitor color coding

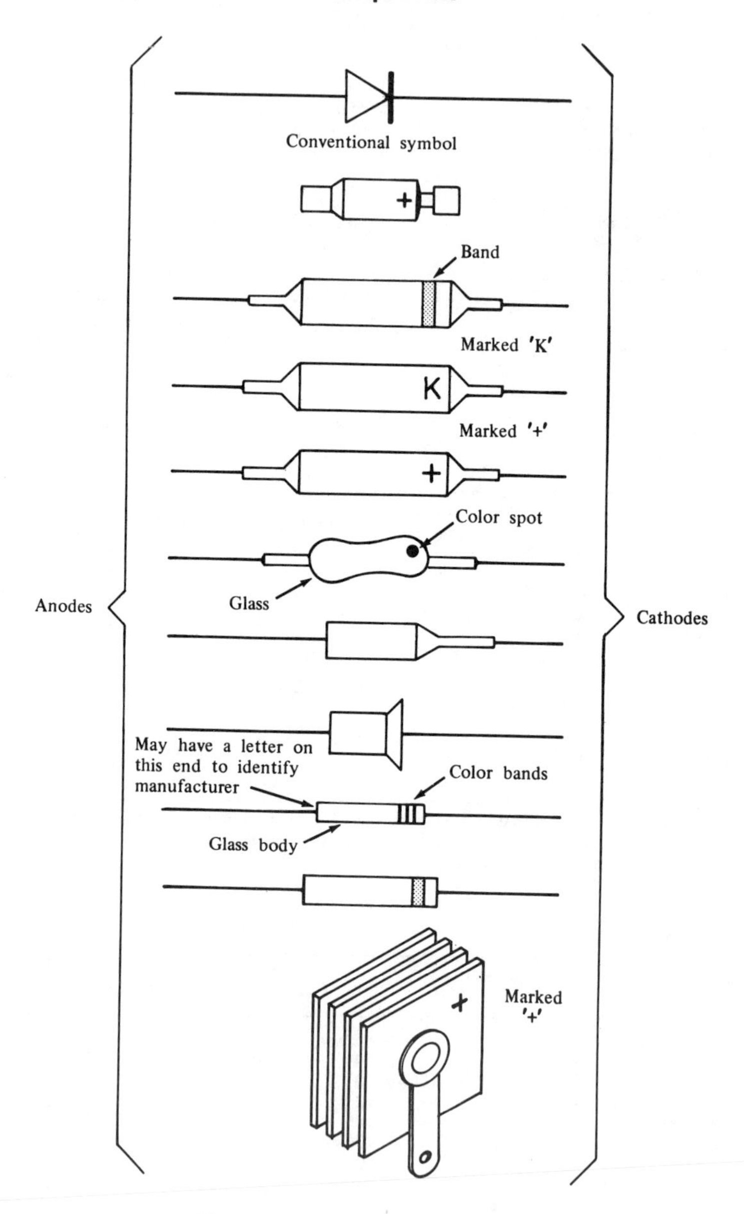

Figure 3-12 Diode color coding

3.4 TRANSFORMER COLOR CODING (RETMA STANDARD)

Some transformer leads are color coded. The colors of the insulation on the leads indicate the connections within the transformer (Figure 3-13).

Transformer color coding (RETMA Standard)

Color of lead	Power Transformer	A - F Transformer (also line-to-grid and tube-to-line)	I - F Trans.	Loudspeaker	
				Field coil	Voice coil
Black	Primary (common for tapped primary	Grid return	Grid or diode return		Start
Black and red	Finish of tapped primary			Start	
Black and yellow	Primary tap				
Red	High voltage	B+	B+		
Red and yellow	High-voltage tap			Finish	
Yellow	Rectifier filament (C.T.-Yellow and blue)	Grid or center-tapped secondary			
Green	Filament NO.1 (C.T.-green and yellow)	Grid	Grid or diode		Finish
Brown	Filament NO.2 (C.T.-brown and yellow)	Plate or center-tapped primary			
Slate	Filament NO.3 (C.T.-slate and yellow)			Tap-slate and red	
Blue		Plate	Plate		

Figure 3-13 Transformer color coding (RETMA Standard)

KT'S KEY TERMS TO REMEMBER

Cathode The electrode that is the primary source of electrons; the end of a diode to which the arrow points.

Ceramic A claylike material that acts as an insulator.

Code A system of characters and rules used to provide information. A "color code," for example, provides information based upon color and its location.

JAN Joint Army-Navy coding methods.

pF Picofarad; the p is a multiplier that equals 1×10^{-12} or means to move the decimal 12 places to the left of the whole number. Thus, one pF equals 10^{-12} farad.

Precision The quality of being exacting or conforming closely to an indicated value.

Prefix A word element attached to the beginning of a word to produce a derivative or another form of the word.

Reliability The measure of stability and dependability; that is, the length of time a piece of equipment will last.

RETMA Abbreviation for Radio-Electronics-Television Manufacturers Association, now the Electronic Industries Association (EIA).

Schematic A diagram of an electrical system or part of it with components symbolically represented.

Significant Figure A number expressing the result of a measurement in which only the last digit is in doubt; for example, 2 has one significant figure, 2.14 has three significant figures.

Tolerance An allowable deviation from a specified value; in components, the permissible limits above and below the coded value.

SELF TEST—COLOR CODES

Fill in the blanks with the resistance values (in ohms) and the tolerances for the codes in the left-hand column. Refer to Figures 3-4, 3-5, 3-6, and 3-7 for assistance.

1. Orange-orange-orange-gold	**1.** ________________
2. Red-red-orange-silver	**2.** ________________
3. Red-red-green-silver	**3.** ________________
4. Red-red-green-silver	**4.** ________________
5. Red-green-red-red-brown	**5.** ________________
6. Orange-red-red-gold	**6.** ________________
7. Yellow-violet-red	**7.** ________________
8. Yellow-black-violet	**8.** ________________
9. Violet-yellow-red	**9.** ________________
10. Yellow-violet-yellow-silver	**10.** ________________
11. Violet-yellow-yellow-red-brown	**11.** ________________
12. Violet-black-yellow-yellow-red	**12.** ________________
13. Blue-brown-red	**13.** ________________
14. Blue-red-brown-silver	**14.** ________________
15. Brown-brown-red-gold	**15.** ________________
16. Gray-black-orange	**16.** ________________
17. F3342	**17.** ________________
18. F2723	**18.** ________________

Fill in the blanks with the color combinations that indicate the resistances in the left-hand column.

19. 2500 ±5% **19.** ____________
20. 570,000 ±10% **20.** ____________
21. 280 ±1% **21.** ____________
22. 75 ±20% **22.** ____________

SELF TEST—CAPACITORS

Referring to Figure 3-9, fill in the blanks with the values of the codes in the left-hand column.

1. Black-red-red-green-black-red **1.** ____________
2. White-red-violet-orange-black-green **2.** ____________
3. Brown-red-brown-red-red-orange **3.** ____________
4. Red-black-brown-green-black-black **4.** ____________

Refering to Figure 3-11, fill in the blanks with the values of the capacitor codes in the lefthand column.

5. (Black body) red-violet-orange-green **5.** ____________
6. Silver-brown-red-red-green-black **6.** ____________
7. (Tubular) brown-red-brown-green-green **7.** ____________

PROJECT: READING COLOR CODES

Materials

A quantity of numbered parts on which the values are identified by color code.

Procedure

List the value of each part supplied by your instructor as in the following example.

Example: Resistor—red, red, orange, silver—22,000 ± 10% ½ watt.

4: Hand Tools

4.0 INTRODUCTION

Electronic assembly is the process of fabricating functional electronic units by properly mounting, connecting, and *encasing* components that are already manufactured. This requires a thorough knowledge of components, materials, and equipment; however, to transform this knowledge into a meaningful project also requires a complete familiarity with the hand tools required to perform the task. Pictures, names, descriptions, and applications of the most commonly used tools are discussed in this chapter. Remember to learn the KT's at the end of the chapter.

4.1 HAND-TOOL SAFETY

The ability to use hand tools effectively depends to a large extent on the degree of care and safety applied when using them. Before we study tool application, let us review some safety rules we must follow.

1. Before working on an electrical unit, turn off all *power* and *short circuit* to ground all *high-voltage points.* A capacitor can store a *lethal charge.* Make sure that the power cannot be *restored* accidentally.
2. When you cut wires, keep the open side of the cutter away from your body and keep cut-off wire ends in your own area.
3. Keep a protective cover over knife blades.

4. Keep tools and materials from *projecting* over the edge of benches.
5. Place tools and materials so they cannot slide, roll, or fall.
6. Never work on rotating machinery with someone standing close by.
7. Never use a box, barrel, chair, or any other makeshift support as a substitute for a ladder.
8. Replace or repair splintered, broken, rough, or loose tool handles before use. Use only files that are equipped with handles.
9. Shield the dangerous parts of sharp-edged or pointed tools when you carry them from place to place.
10. Keep screwdriver blades sharp and even. (Do not use screwdrivers with broken or rounded points or broken handles.)
11. If a grinding wheel vibrates or wobbles excessively, stop it at once and report the condition to the instructor or shop supervisor.
12. When using a drill press, hold the work with a *drill vise* or some other *clamping tool.*

4.2 PLIERS

There are numerous types and styles of pliers, the general purposes and unique applications of which vary with the type and style. The following pliers are those most often used in electronic assembly.

Long needle-nose pliers have long, slender jaws terminating in a fine point (Figure 4-1). They are fragile and intended for delicate handling of small parts and small *diameter* wire and for reaching into hard to get at places. They are never used for bending wire or as a wrench.

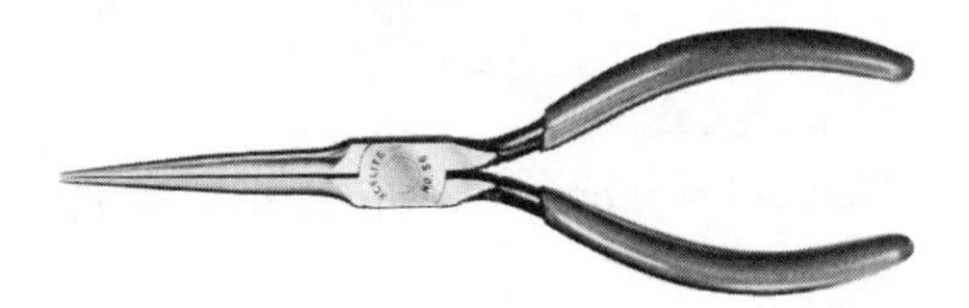

Figure 4-1 Long needle-nose pliers

Slim-nose pliers are similar to the needle nose but the long tapered jaws are heavier at the base for added strength (Figure 4-2). These pliers are used for bending and forming small, thin wire and for holding small parts. They should not be used for bending heavy wire.

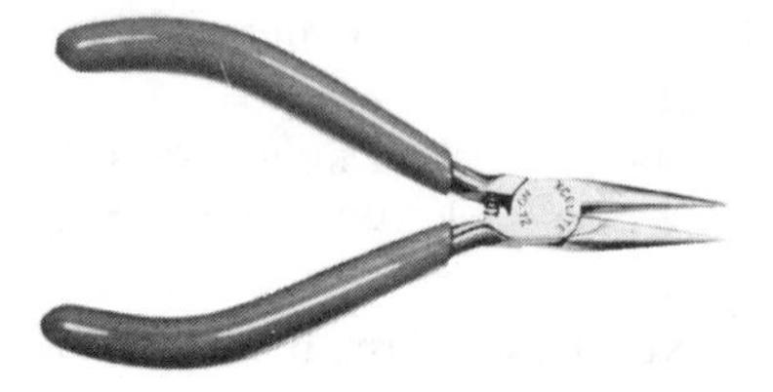

Figure 4-2 Slim-nose pliers

Needle-nose pliers have a very fine tip and slim jaws, but the base of the slim jaws is anchored to heavier jaws which add firmness to the needle point (Figure 4-3). The needle-nose plier is used for handling small wire and parts. It provides approximately the same tip as the long needle nose but with a firmer jaw. When needed, a firmer needle-nose grip can be obtained.

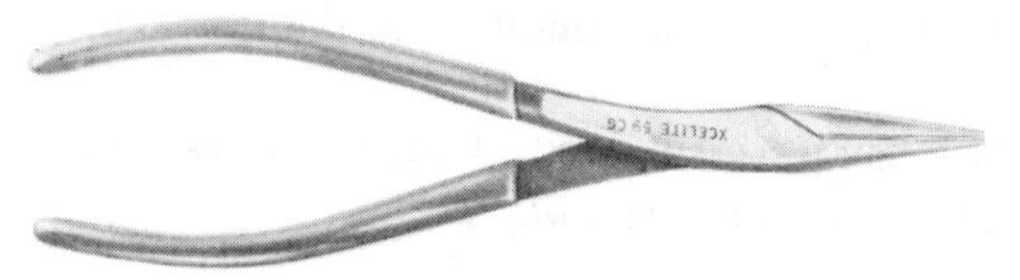

Figure 4-3 Needle-nose pliers

Long-nose tip-cutting pliers have a firm jaw much like the slim-nose plier, but they also have the added feature of a wire cutter near the tip (Figure 4-4). This enables cutting wire in hard to get at places. Because the tip extends beyond the cutter, this plier can also be used as a long nose.

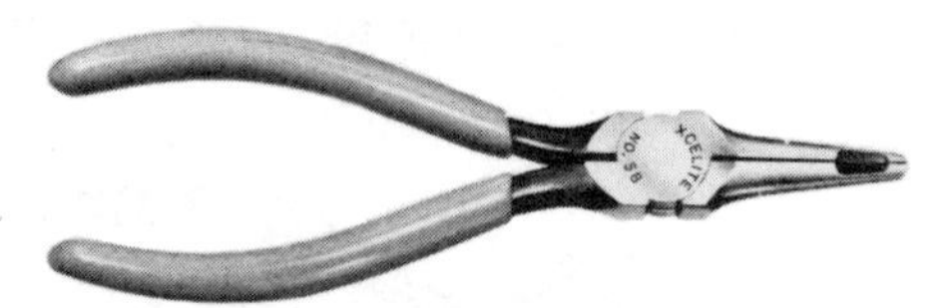

Figure 4-4 Long-nose tip-cutting pliers

Diagonal cutting pliers are designed with a short nose and flush cutting edges (Figure 4-5). The small, close to the joint jaws make this cutter easy to handle without using brute strength. It should be used for cutting small wire only, if any cutting requires more strength than a normal hand grip, this plier should not be used. When a strong grip is required, use a heavy diagonal plier.

Figure 4-5 Diagonal cutting pliers

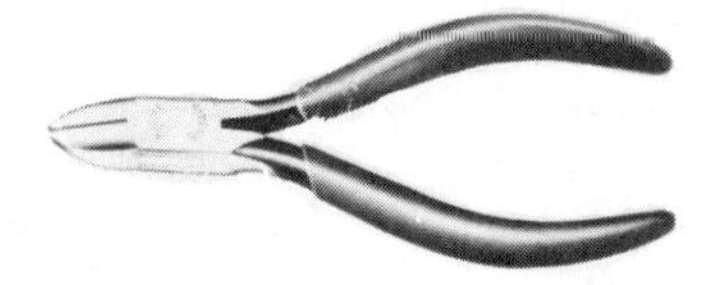

Short-nose heavy diagonal pliers have a short nose and a heavy jaw with side cutting edges (Figure 4-6). The extra large handles and short cutting edges make this a powerful cutting tool ideal for a heavy wire. It can also be used to trim printed-circuit boards and plastics used in assembly.

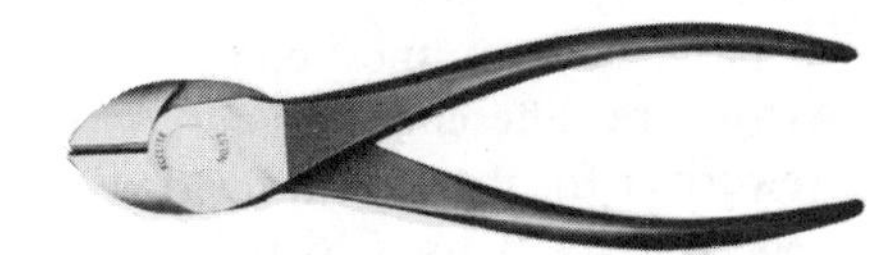

Figure 4-6 Short-nose heavy diagonal pliers

Linesman pliers, also known as electricians' pliers, have strong, *knurled* jaws and a heavy-duty side cutter (Figure 4-7). They are designed for heavy-duty work and the side cutters, as well as the jaws, will serve for most of the heaviest work encountered in assembly.

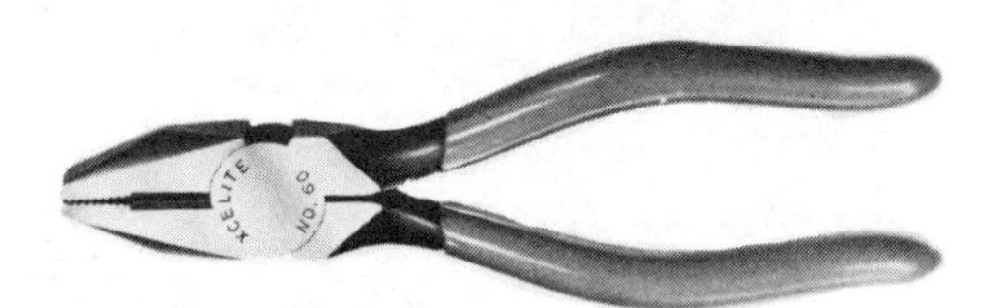

Figure 4-7 Linesman pliers

The dual-blade wire stripper is one of several types of wire strippers (Figure 4-8). Most types are similar to pliers in design but differ by having cutting edges notched to different wire sizes. This tool enables you to strip off insulation without nicking or cutting the wire. Care must be taken to insert the wire to be stripped into the correct notch.

4.3 SCREWDRIVERS

Screwdrivers are among the most commonly used hand tools in the trade. They generally consist of two parts: the blade and shaft, and the

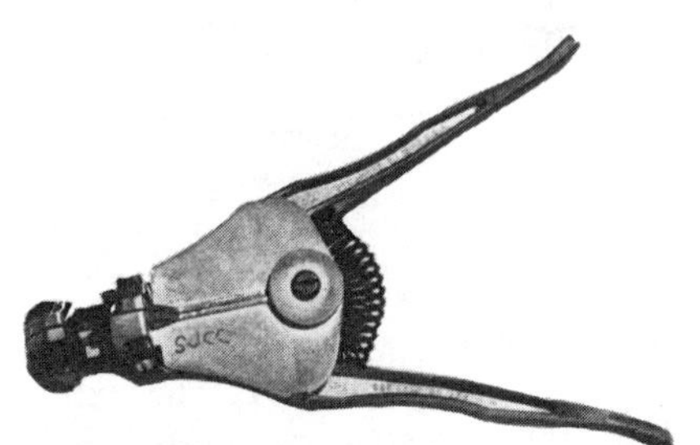

Figure 4-8 Dual-blade wire stripper

handle, which is permanently attached. Even though they are manufactured in only two basic designs, they come in many sizes because different sized screws require different sized screwdrivers. Selecting the appropriate sized screwdriver for the task at hand is essential to success.

The flat-blade screwdriver is used to turn a screw with a slotted head (Figure 4-9). The slot in the screw head creates two walls which press against the edges of the screwdriver tip. In order to avoid any play, which would ruin the screw head, a properly fitting screwdriver should be used at all times. Remember, the larger the screw, the larger the slot, therefore, the larger the screwdriver.

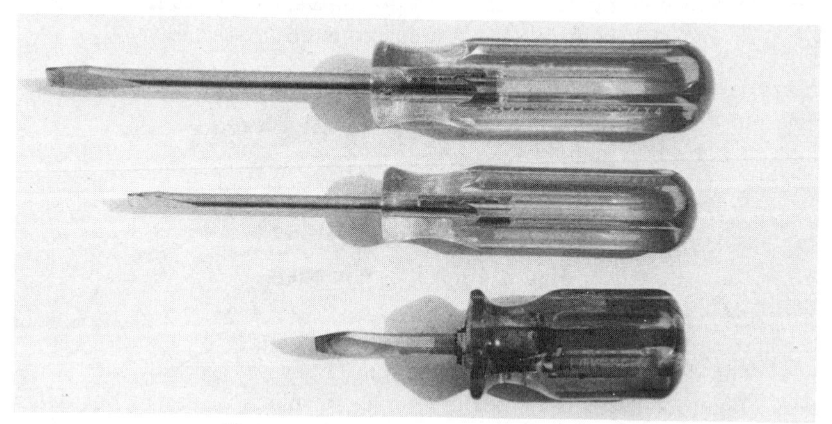

Figure 4-9 Flat-blade screwdrivers

The Phillips screwdriver has a notched point that, when fitted properly into a screw head, applies pressure on four walls of the screwhead (Figure 4-10). The Phillips, like the flat-blade, comes in varying sizes, so care should be taken to select the appropriate sized screwdriver point. The correct size of screwdriver will fill the notched hole in the screw head and apply pressure on all four walls simultaneously.

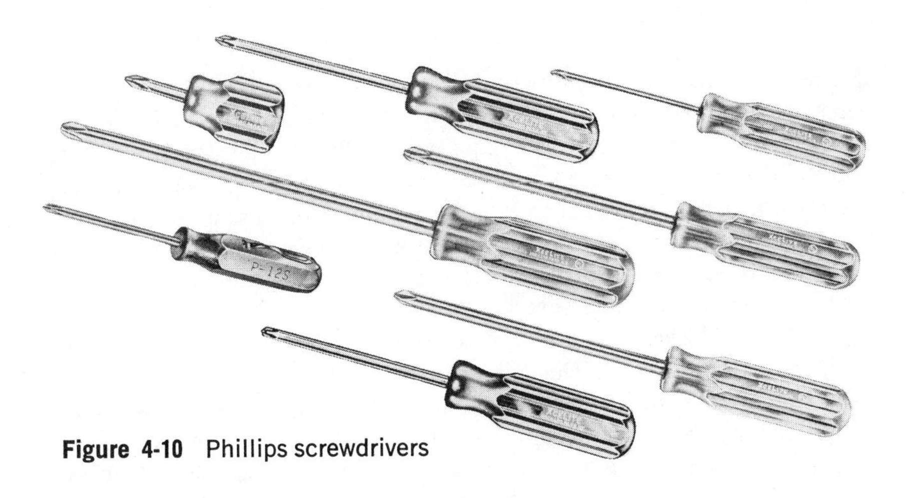

Figure 4-10 Phillips screwdrivers

4.4 NUT DRIVERS

The nut driver, another commonly used tool, is a fixed socket wrench (Figure 4-11). It comes in sets, the most often used sizes of which range from 3/16 to 3/8 in. The nut driver is similar to the screwdriver in design except that it has a socket at the end instead of a point. The thickness of the socket walls varies with the size of the socket. The shaft may vary in length but the most often used length is approximately 4 in. The nut driver is a very useful tool for quickly removing or tightening nuts or lugs. It is handled like a screwdriver, that is, turned with one hand. Care should be taken not to overtighten a nut or lug since extra *leverage* is provided by the large handle.

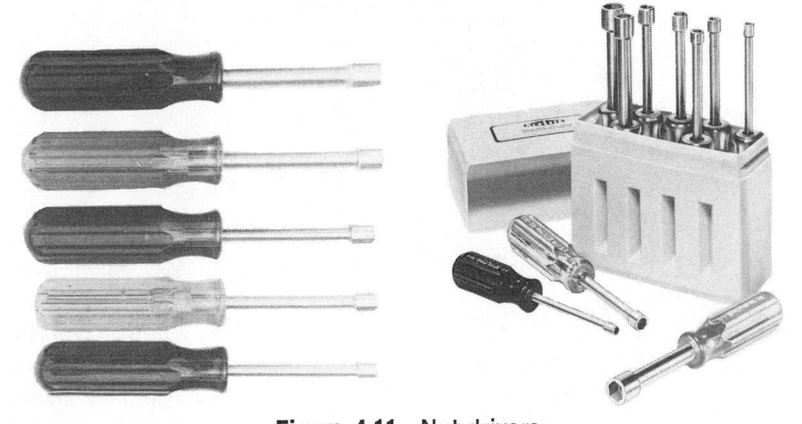

Figure 4-11 Nut drivers

4.5 SOLDERING TOOLS

Using the soldering iron properly is one of the most demanding skills of an electronics assembler. The soldering iron, like other tools, has to be selected to fit the job. It also requires careful handling from a safety point of view as it can cause severe burns. Furthermore, incorrect use of a soldering iron can cause a faulty soldering operation. A unit may be inoperative because a proper-appearing solder is in fact a poor one (cold solder joint). Care of soldering irons is discussed in Chapter 7. Here we will discuss and describe them.

The soldering iron shown in Figure 4-12 has a copper tip that is heated by a heating coil. The coil is connected to the ac power plug from which it receives power for heating and will maintain a temperature of approximately 600° at the tip. The purpose of the soldering iron is to transfer enough heat to the parts you want to join to melt the solder but not the parts. When the parts being joined are hot enough to melt the solder, the solder is applied to the parts. As soon as the solder is

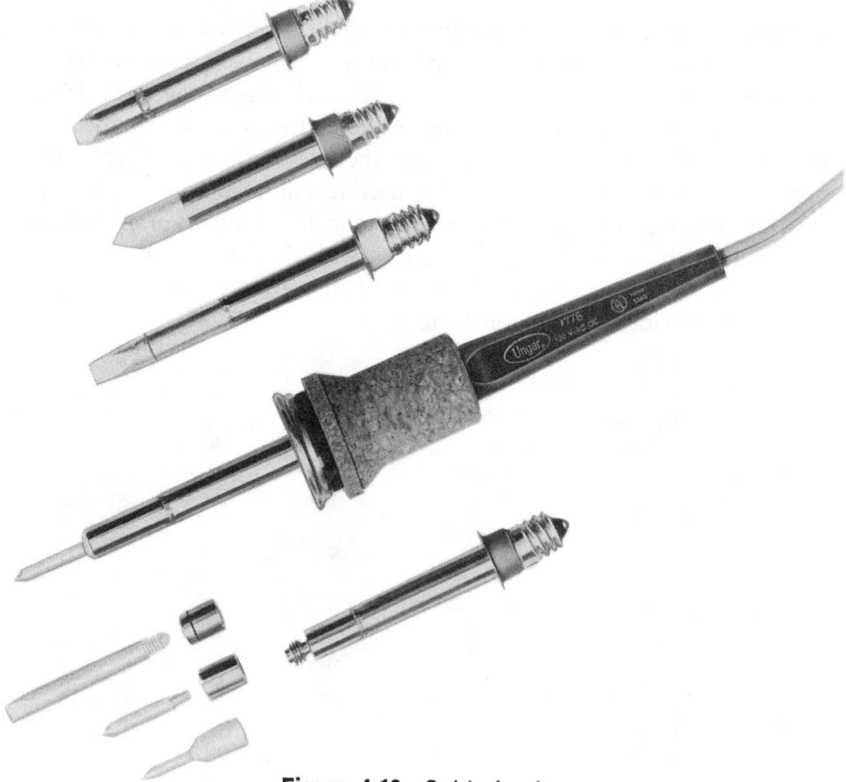

Figure 4-12 Soldering iron

melted, the iron is removed and the parts are permitted to cool without movement. A properly soldered joint will make a good electrical connection but soldering is not used to provide strength to the joint.

Remember—the soldering iron gets hot and can cause severe burns; therefore, hold it only by the handle.

The antiwicking tool (Figure 4-13) works like a self-closing plier—when the handle is released, the jaws tighten. An antiwicking tool prevents solder from running under the insulation when tinning a wire. It can also be used as a *heat sink*.

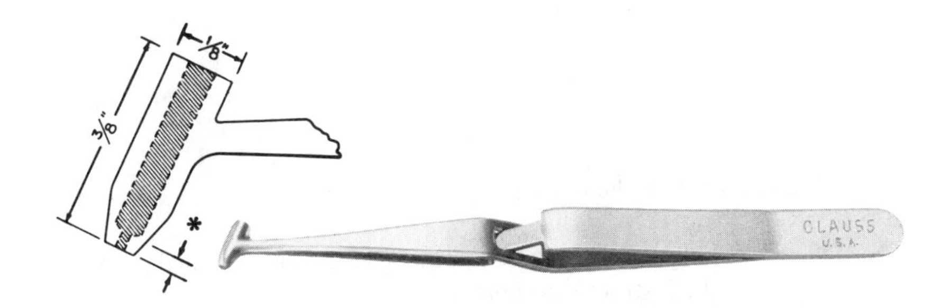

A soldering aid is a handy tool with applications in most soldering jobs (Figure 4-14). One end is pointed and used to unwind wires and clean holes before soldering. The other end can be used to twist or untangle hook-up wire or to hold parts while soldering. The two ends are connected by an insulating material such as wood that serves as a handle.

Figure 4-14 Soldering aid

KT'S KEY TERMS TO REMEMBER

Clamping Tool A tool used to hold a piece that is being drilled.
Diameter The thickness of a material such as wire, given in inches, centimeters, or mils (0.001 in.).
Drill Vise A vise used to clamp objects to be drilled.
Encasing Enclosure or covering.
Heat Sink A tool or device used to absorb or dissipate unwanted heat.
High-Voltage Point A point that carries a dangerous voltage.

Knurled Ridged to help grip an object.
Lethal Capable of causing death.
Leverage Lifting force; a wrench with a long handle provides added force at the nut.
Power The source of energy to a circuit.
Projecting Extending or sticking over or out from something.
Restored Returned to its initial value.
Short Circuit An abnormal connection between two points resulting in a flow of excess (and often undesirable) current.

SAFETY TEST

Fill in the blanks with the correct word or words.

1. Before working on electrical equipment you should turn the power off and ask the instructor to help you ______________ all high-voltage points.
2. A capacitor can ______________ a lethal charge of electricity.
3. Make sure the ______________ cannot be accidentally restored to a piece of equipment being repaired.
4. You should cut off wire ends so that they remain in ______________ area.
5. The ______________ side of the cutter should be kept away from your body.
6. Knife blades should have a ______________ over them.
7. Make sure people are not ______________ to rotating machinery you plan to use.
8. To reach above the head you should use a ______________, not a chair or box.
9. Files must be equipped with ______________ before use.
10. Never use tools that have broken ______________.
11. If you must carry sharp or pointed tools, they should be ______________.
12. If a grinding wheel wobbles or vibrates, you should immediately ______________ the instructor.
13. When using the drill press, hold the work with a drill press ______________ or a ______________ tool.

HAND-TOOL TEST

Match the name to the description by placing the number of the description in the blank opposite the name of the appropriate tool.

Name of Tool	*No. of Description*
1. Dual-blade wire stripper	__________
2. Long needle-nose pliers	__________
3. Slim-nose pliers	__________
4. Long-nose tip-cutting pliers	__________
5. Diagonal cutting pliers	__________
6. Linesman pliers	__________
7. Flat-blade screwdriver	__________
8. Phillips screwdriver	__________
9. Nut driver	__________
10. Soldering iron	__________
11. Antiwicking tool	__________
12. Soldering aid	__________

Hand-Tool Description	*Number*
The cutter near the tip of these pliers helps cut wire in hard to get at places.	
These delicate pliers are designed to cut without the need for brute strength.	
These pliers have a knurled jaw and can be used for heavy cutting.	
This device contains a heating coil, a copper tip, and a handle.	
This tool is used to prevent solder from running under the insulation.	
This tool has one pointed end, one with two prongs, and a center made of an insulating material.	
This tool has a blade that fits into a screw slot.	
A fixed socket wrench that comes in a set.	
A round rod with a notched point that can apply pressure on four walls of a screw slot.	
A tool that looks like a pair of pliers with notched cutting edges.	
Very fragile pliers for delicate handling of small parts.	
Very delicate pliers used to form small wires.	

PROJECT: IDENTIFYING TOOLS

Materials

Box of numbered hand tools typical of electronics assemblers.

Procedure

Identify each of the tools furnished by your instructor by name and function as in the following example.

Example: Phillips screwdriver—used to tighten or loosen Phillips type screws.

5: Hardware

5.0 INTRODUCTION

Hardware is the name given to all those parts used to fasten the components of a typical assembly. Even though they are the parts that cost least, they determine the steadiness and physical durability of most assemblies. Hardware is commonly *classified* as screws, washers, nuts, clamps, brackets, and fasteners. This chapter describes and lists the applications of these categories. Remember to learn the KT's at the end of the chapter.

5.1 SCREWS

Screws are available in many shapes and sizes with many head designs to meet specific needs. The most commonly used screw is the machine screw. Following is a categorization of machine screws by head design and application.

The round-head screw shown in Figure 5-1 is a general purpose screw that is used more often than any other type. This screw is often used where one of the types presented later in the chapter should be used.

The pan-head screw (Figure 5-2), which is available both with Phillips- and flat-blade screwdriver type heads, has replaced the old round-head screw because it provides greater strength at the edges of the screw head. The pan-head screw's rugged design provides holding power with little danger of ruining the screw head while it is being tightened.

The fillister screw head (Figure 5-3) is smaller than the pan head with deeper slots that provide a good hold for the screwdriver. Varieties are made for both Phillips and flat-blade screwdrivers. The smaller diameter of the fillister screw head increases the pressure applied to the part being held. It is used when room is limited, such as when the screw hole is located close to a raised surface, because the high head keeps the screwdriver away from the surface area. It is often used in counter-bored holes to provide a flat surface.

The binding-head screw's (Figure 5-4) unique feature is its recessed undercut which provides the screw with pressure around the edge rather than at the center. For this reason it will apply pressure and bind but will not fray a strand-type wire. It is often used in radio or electrical type work. For this purpose it is not usually found as a Phillips head.

The *truss* or oven-head screw (Figure 5-5) has a smooth, almost flat head that will blend into a surface. Its unique application is to fill or cover an extra large hole while still providing some holding power, thus, it is a strong finishing screw.

The oval-head screw (Figure 5-6) design is very much like the round head, the predecessor of the pan-head screw. However, this screw has a beveled head while the pan and old round head are square sided. It is available in either a Phillips or slot-type head. The oval head is nearly as strong as the pan head but has a smoother appearance and will fit a *countersink*. Oval heads are used to help hold components in critical positions.

The flat-head; countersunk screw (Figure 5-7) has a small head with an 80 to 82° *angle*. It is available in both Phillips or slot-type heads. The flat-head countersunk screw is used when a flat surface is desired because, if properly countersunk, it will remain *flush* with the surface. It is very effective for centering or maintaining positioning of parts.

The flat-head screw (Figure 5-8) has the same design as the 80 to 82° angle flat head except that it has a 100° angle to its head. It also fits flush to create a smooth surface but it can spread pressure over a greater area than its 80° counterpart. For this reason, it is useful for soft surfaces.

The washer-head screw (Figure 5-9) has a round head on which a flange has been left to provide greater pressure on the surface at the edges. This screw is normally used when a larger surface is required than provided by the pan or round head and a washer would be too difficult to *manipulate*.

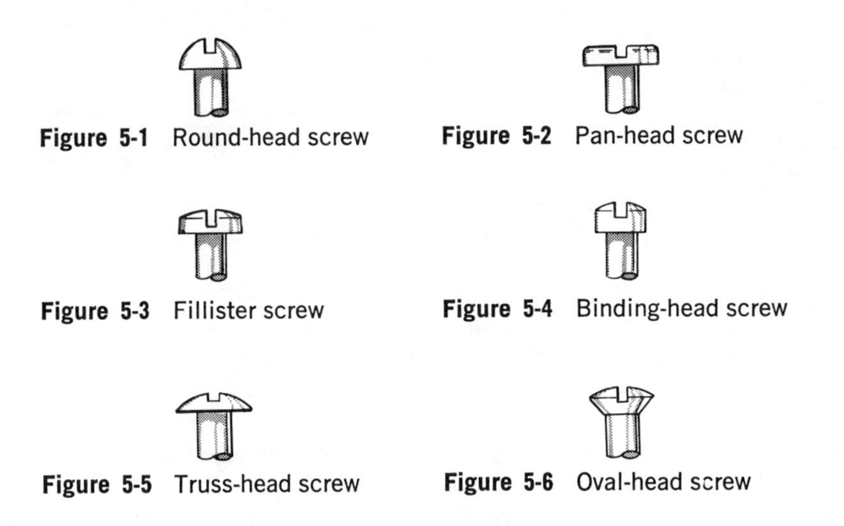

Figure 5-1 Round-head screw

Figure 5-2 Pan-head screw

Figure 5-3 Fillister screw

Figure 5-4 Binding-head screw

Figure 5-5 Truss-head screw

Figure 5-6 Oval-head screw

Figure 5-7 Flat-head countersunk screw

Figure 5-8 A 100° flat-head screw used for soft surfaces

Figure 5-9 Washer-head screw

The hexagon-head screw (Figure 5-10 a), also known as a bolt, has a six-sided head with sharply defined corners. It is available in many sizes with the head size varying according to the threaded bolt diameter. The hexagon-head screw is used when more than average pressure must be applied; thus it is turned by means of a wrench which permits much greater leverage or *torque* than a screwdriver. A variety of wrenches—open-end, box-end, or socket [Figure 5-10 (b) to (d)]—may be used.

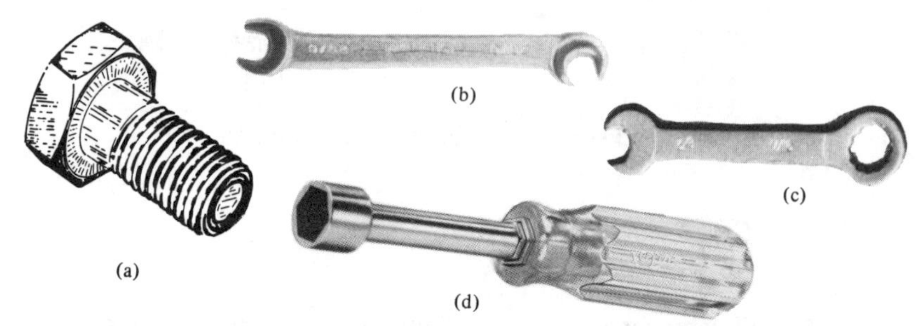

Figure 5-10 Hexagon-head screw (bolt) and wrenches: **(a)** screw, **(b)** open-end wrench, **(c)** box-end wrench, **(d)** socket wrench

Tapping screws (Figure 5-11), which are designed to cut their own threads, are used on sheet metal and other similar materials. There are several designs differing principally in thread style and point.

Set screws (Figure 5-12) are small in diameter and designed to be turned by a wrench or small-blade screwdriver. The set screw is used to hold shafts and keep them from turning. Selecting the type of set screw is determined largely by the pressure required and the freedom to leave a mark on the shaft.

A summary of screw-head types and drives is shown in Figure 5-13(a). Examples of set screw tips are shown in Figure 5-12(b). The Allen wrench set shown in Figure 5-12(c) is typical of wrenches used to tighten Allen-set screws.

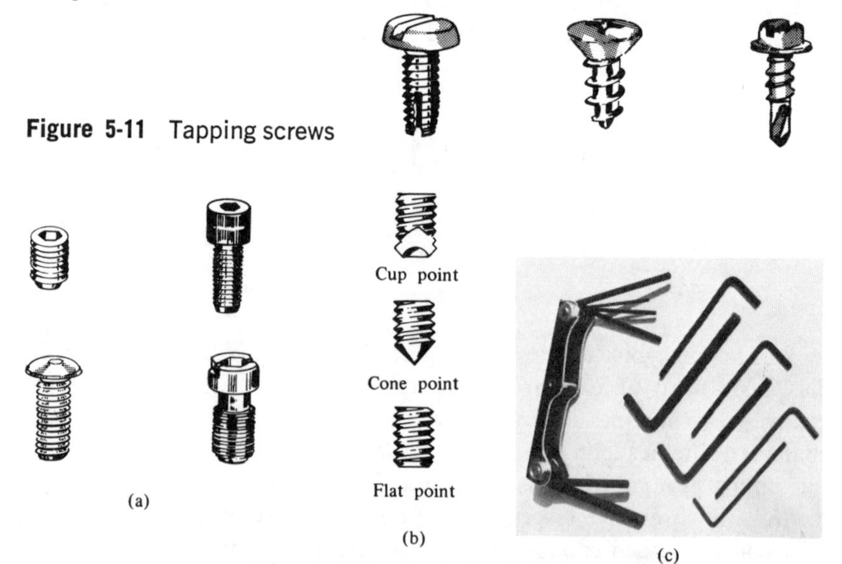

Figure 5-11 Tapping screws

Figure 5-12 Socket-set screws

Spline Socket for High - Torquing Screws

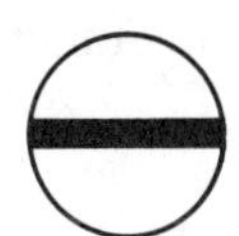

Slotted

This is the traditional screwdriving means - a center slot accepting the standard blade screwdriver.

Pozidriv®

This is the newest development in a recess. This design provides a positive fit to eliminate wobble or cam out.

Phillips

This is the more conventional crossed recess designed to provide good control in driving.

Frearson

Sharp - cornered cross recess gives close fit and good control.

Clutch

This recess provides straight sides and positive fit for bit and head.

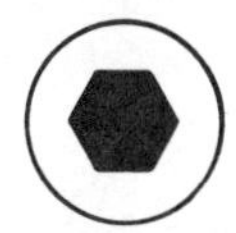

Hex Socket

Socket recesses provide an effective means for driving close tolerance products in areas with no clearance.

Drilled Spanner

Holes provide means for driving or removing this head with special tool. Prevents unauthorized tampering.

Slotted Spanner

Notches on circumference accommodate special driving tool. Same purpose as Drilled Spanner.

One - Way

This is a basic slot with the reverse side cammed so that screwdriver can only advance screw and not back it.

Figure 5-13 A summary of screw-head drives and styles *(Courtesy of Sloss Fasteners)*

There are many types of nuts (Figure 5-14), all designed for specific applications. Their common design characteristic is the ability to be turned either by hand or with a wrench. Nuts apply pressure between the head of the screw or bolt and the object being held. When tightening a nut with a wrench, you must be careful not to round the nut or *strip* the *threads*.

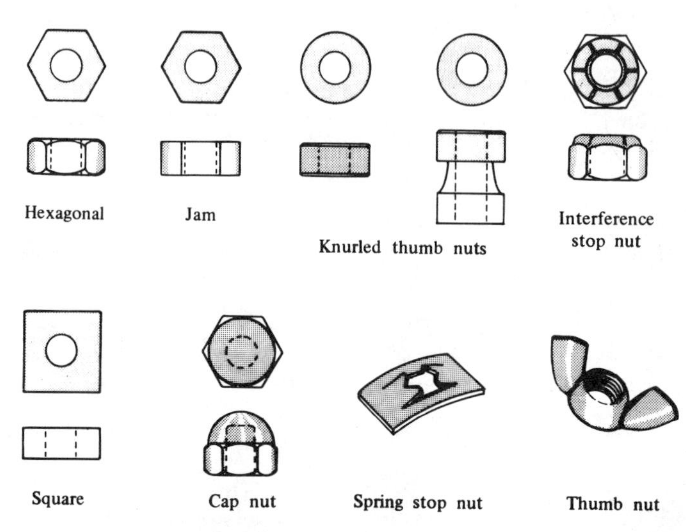

Figure 5-14 A summary of nuts

5.2 WASHERS

Washers (Figure 5-15) are round, flat disks with center holes that come in a number of variations to meet specific needs. They may be made of either metal or nonmetallic material as the need dictates. There are three basic categories of washers: (1) The flat washer protects the surface from the turning nut or screw. (2) The spring lock washer protects the surface also, but remains under spring *tension* when tightened. This keeps the nut from *vibrating* loose. (3) The *external* and *internal* teeth *binding* lock washers are used when a metal to metal contact is necessary. The teeth cut through a painted or varnished surface.

5.3 BRACKETS AND CLAMPS

Brackets and clamps (Figure 5-16) are found in a variety of shapes and sizes but the main difference, besides size and basic design, is the

External Type

External type lock washers provide greater torsional resistance due to teeth being on largest radius. Screw heads should be large enough to cover washer teeth. Available with left hand or alternate twisted teeth.

Internal Type

For use small screw heads or in applications where it is necessary to hide washer teeth for appearance or snag prevention.

External - Internal Type

For use where a larger bearing surface is needed such as extra large screw heads or between two large surfaces. More biting teeth for greater locking power. Excellent for oversize or elongated screw holes.

Heavy Duty Internal Type

Recommend for use with larger screws and bolts on heavy machinary and equipment.

Dome Type Plain Periphery

For use with soft or thin materials to distribute holding force over large area. Used also for oversize or elongated holes. Plain periphery is recommended to prevent surface marring.

Dome Type Toothed Periphery

For use with soft or thin materials to distribute holding force over large area. Used also for oversize or elongated holes. Toothed periphery should be used where additional protection against shifting is required.

Countersunk Type

Countersunk washers are used with either flat or oval head screws in recessed countersunk applications. Available for 82° and 100° heads and also internal or external teeth.

Dished Type Plain Periphery

Recommended for the same general applications as the dome type washers but should be used where more flexibility rather than rigidity is desired. Plain periphery for reduced marring action on surfaces.

Figure 5-15 A summary of washer types *(Courtesy of Sloss Fasteners)*

Dished Type Toothed Periphery

Recommended for the same general applications as the dome type washers but should be used where more flexibility rather than rigidity is desired. Toothed periphery offers additional protection against shifting.

Pyramidal Type

Specially designed for situations requiring very high tightening torque. The pyramidal washer offers bolt locking teeth and rigidity yet is flexible under heavy loads. Available in both square and hexagonal design.

Finish Type

Recommended where marring or tearing of surface material by turning screw head must be prevented and for decorative use.

Helical Spring Lock Type

Spring lock washers may be used to eliminate annoying rattles and provide tension at fastening points.

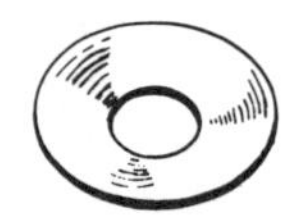

Cone Spring Type

Plain periphery. Referred to as "Belleville" washer. A spring action washer used to distribute holding force over larger area. For oversize or elongated holes. Takes high tightening torque.

Cone spring type serrated periphery

Same general usage as the cone type with plain periphery but with the added locking action of a serrated periphery. Takes high tightening torque.

Flat Type

For use with oversize and elongated screw holes. Spreads holding force over a large area. Use also as a spacer. Available in all metals.

Fiber and Asbestos

In cases where insulation or corrosion resistance is more important than strength, fiber or asbestos washers are available.

Figure 5-15 (continued)

covering of the clamp or bracket. For example, some are coated with plastic or rubber to protect surfaces. Clamps and brackets are used to anchor or hold wires, cables, parts, and subassemblies to a fixed location. They provide a physical attachment that reduces movement and strain on electrical connections.

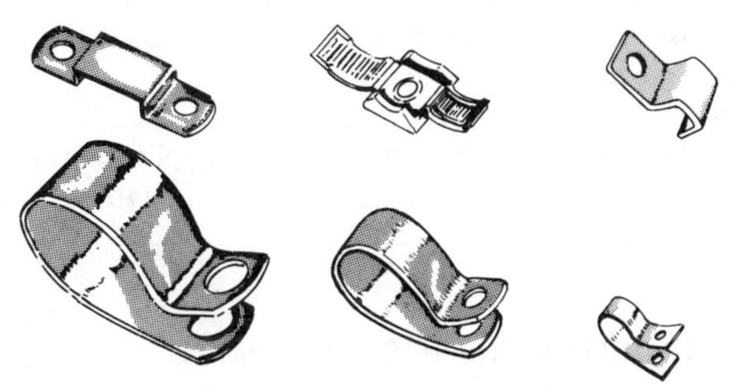

Figure 5-16 Brackets and clamps

KT's KEY TERMS TO REMEMBER

Angle The figure formed when two surfaces extend from the same line.
Bevel To cut or shape a surface; a slanted or angled surface.
Binding The action of fastening together.
Classified Arranged in certain categories.
Countersink To set a head at or below the surface of a material; a funnel-shaped enlargement of the outer edge of a hole.
Flush Smooth to a surface.
Manipulate To handle or control.
Torque A turning or twisting force.
Truss A binding force.
Vibrating Moving side to side, shaking.

TEST

Fill in the blanks.

1. Electronic hardware is defined as ______________________.
2. If you need a screw with holding power and a strong screw head, you would ask for a ______________________ screw.
3. If you need to hold a wire under the screw head, you would use a ______________________ head screw.

4. You request oval- or flat-head screws when a ______________ ___________ screw is required to do the job.
5. If you want a flat surface finish you would use a ____________ ___________ head ____________________ screw.
6. A tapping screw is used for ____________________.
7. A screw with a hexagon head is known as a ________________.
8. A bolt is tightened with a ________________________ which provides greater leverage than a screwdriver.
9. A socket-set screw is used to ____________________.
10. Three different types of washers and their applications are _______ _________________, _________________, and ____________ ____________________________________.

PROJECT: IDENTIFYING HARDWARE PARTS

Materials

A quantity of typical hardware parts with each part numbered.

Procedure

List the name and function of each of the numbered hardware parts as in the following example.

Example: Tapping screw—designed to cut its own threads in sheet metal and other similar materials.

6: Parts Mounting

6.0 INTRODUCTION

Producing most manufactured products requires mounting parts. Parts are mounted at different stages of production; how well they are mounted largely determines the product's *durability*. Loose or carelessly mounted parts will result in an *inoperative* or even dangerous piece of equipment. Figures 6-1 through 6-3 *illustrate* basic hardware stackups.

The *ground-lug* stackup in Figure 6-1(a) has an external tooth *lock washer*. The lock washer bites into the panel creating a good electrical connection. This stackup is tightened by turning the nut, not the screw.

The stackup in Figure 6-1(b) has an external screw head and is tightened by turning the screw. This draws the external lock washer into the panel, also creating a good electrical connection.

The hardware stackup in Figure 6-2 uses an external tooth lock washer to make the electrical connection. The whole assembly may be tightened by turning the screw or the nut.

When installing a cable clamp (Figure 6-3), an electrical connection is not needed. The clamp must have a flat metal *eyelet* or a flat washer next to it but tooth washers are not necessary. Figure 6-3(a) shows the clamp being installed on a flat base. Figure 6-3(b) shows the clamp being installed with a *standoff* for *elevation*. Remember to learn the KT's at the end of the chapter.

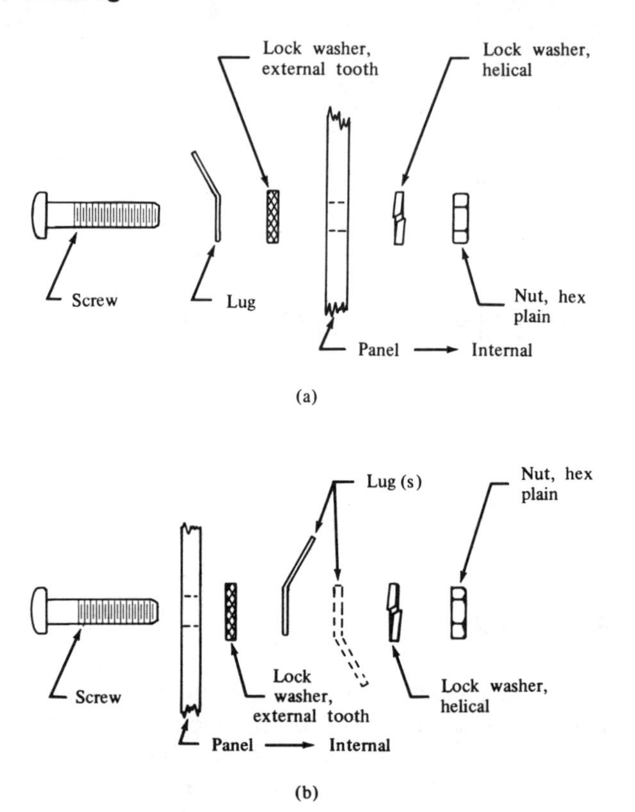

Figure 6-1 Two methods of installing ground-lug stackups

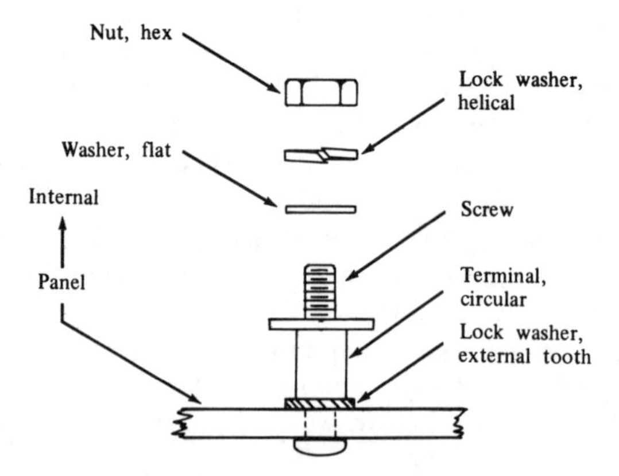

Figure 6-2 Hardware stackup for standoffs

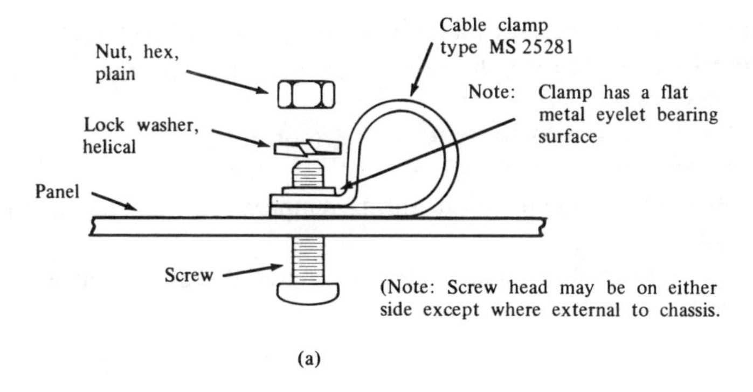

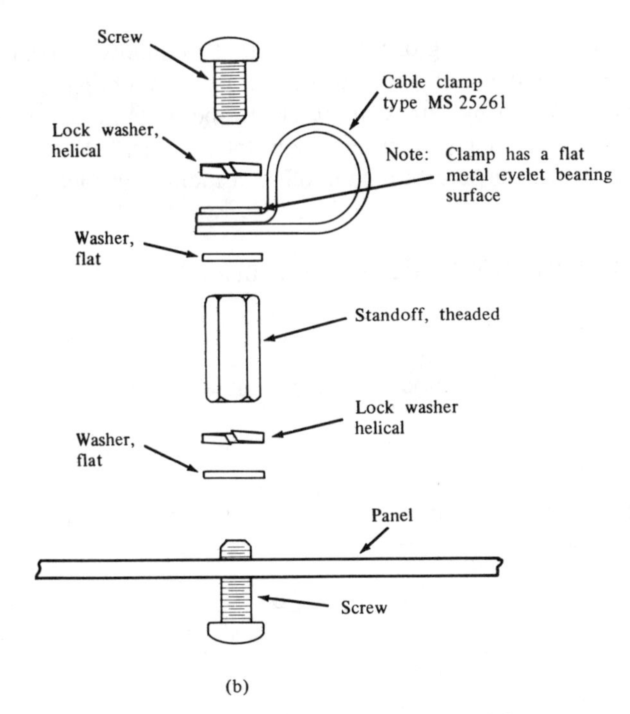

Figure 6-3 Two methods of mounting cable clamps: **(a)** flat base, **(b)** for elevation

KT's KEY TERMS TO REMEMBER

Durability Ability to last a long time without deterioration; for example, how long a part or unit will operate effectively.

Elevation The height above which something is raised.

Eyelet A tubular piece of metal with one end or side rolled over.
Ground Lug A lug for connecting a grounding conductor to a grounding electrode.
Illustrate Show how to do something; demonstrate.
Inoperative Not functioning.
Lock Washer An open, spiral, spring-tempered washer that will lock a nut in place.
Standoff A metal or insulated piece of material that separates wires or parts from the chassis.

TEST

1. Why is the mounting of parts important to quality control?
2. What use is made of a lock washer in parts mounting?
3. What is meant by "stackup" in electronic parts mounting?
4. Does a cable clamp require a tooth lock washer? Why?
5. What is the purpose of a standoff in a mounting stackup?

PROJECT: MOUNTING A CABLE CLAMP

Materials

Parts needed to mount a typical cable clamp.

Procedure

Use the parts supplied by your instructor to properly connect a cable to a chassis.

7: Soldering

7.0 INTRODUCTION

Soldering is the process of joining two metals without melting the two metals being joined. What is melted and what *fuses* to the two metals *molecularly* is the solder. Solder is an *alloy,* generally composed of tin and lead, with a melting point lower than that of the metals to be joined.

When heat is applied to a solder connection and solder is applied, the solder flows around the joint and mixes with the *molecules* on the surface of the metals to be joined. Upon cooling, the connection becomes a continuous metal surface. For example, before two copper wires are soldered together they are two separate wires. They can be tied together mechanically (wrapped or twisted together), but they are still two wires. However, when the joint is heated to the melting point of solder and solder is applied, both a chemical and metallurgical reaction take place. The solder causes a small amount of the copper to dissolve at temperatures well below the melting point of copper, causing the formation of a new alloy that is neither solder nor copper. Upon cooling, a continuous metal connection is formed fusing the two copper wires together so they are no longer two separate wires.

This fusion of the base metals must take place from a good electrical connection. A simple mechanical connection between two wires, or between component leads and wires, is not acceptable in electronics equipment for two reasons: (1) When copper is *exposed* to air, an *oxide* forms on the surface of the copper (like rust on iron). This oxidation is greatly increased by the application of heat and it further *accelerates* with age. Because this oxide coat forms on the surface of copper wires, the wires are really not touching when spliced (physically twisted) together—they are actually separated by the oxide coating. This causes the *conductivity* at the joint to be unstable; that is, the resistance will be higher than if the wire were continuous and will change as the coating of oxide changes. (2) The connection is not solid; the wires will move easily at the joint and may even come apart causing a broken (open) connection.

For these reasons, soldering is a necessary and important operation in electronic circuit construction. Remember to learn the KT's at the end of the chapter.

7.1 BASIC RULES FOR SOLDERING

The following rules should be followed when soldering:

1. Clean the tip of the heated soldering iron by wiping it lightly on a Kimwipe, dry rag, or by using a sponge that has been moistened with water.
2. Prepare the soldering iron by applying *flux* and solder to the clean tip the moment it is hot enough to *liquefy solder*. This is called *tinning* the iron. The tinning action coats the tip with solder and protects it from oxidation. A spot of solder is left on the tip of the iron between uses to help keep the tip clean and prolong tip life.
3. Apply *rosin*-type flux only. When *cored solder* is used, the flux is applied automatically just before the solder liquefies because the flux in the core melts at a lower temperature than the solder. Additional flux is often used for best results.
4. Inspect terminals and wires for cleanliness and clean with solvent or lead cleaner.
5. Place the soldering iron tip on the terminal and immediately melt a small amount of solder at the point of contact. This forms a solder bridge through which the heat can flow more rapidly to the terminal.
6. Attach the wire or wires to the terminal in a manner suitable to the type of terminal used.
7. When the terminal reaches the melting temperature of the solder, apply more solder to the terminal (not the soldering iron). Do this quickly, melting as much as is needed without delay.

8. As soon as enough solder is applied, remove the solder and then the iron.
9. Finally, allow the connection to cool without movement until the solder *solidifies*. After all soldering is completed, the *flux residue* may be removed by cleaning with a solvent. Never scrape or wire brush to remove the flux residue since this would weaken the connection.

These nine points are an overview of soldering. We will now discuss specific applications of soldering in detail.

7.2 TINNING A WIRE

Wires are tinned before being soldered so that solder can flow easily into the connection. The basic tool for most soldering operations is a soldering iron such as shown in Figure 7-1. The tip of the soldering iron in Figure 7-1 is copper for good heat conduction and, as with most soldering irons, is iron coated to prevent pitting. This is necessary because the temperature of the tip may approach *500°F (260°C)*. Solder melts at approximately 300°F (167°C). *This is mentioned to point out that caution must be taken to prevent damage to parts, wiring, and printed-circuit boards due to application of excessive heat. Care must also be taken to prevent bodily burns.*

Remember that, when plugged into an ac line, many soldering irons have a voltage value on the tip with respect to ground. This voltage can harm transistors or other sensitive components. Therefore, it is good practice to use an *isolation transformer* or to connect a *jumper lead* from the metal body of the soldering iron to the *common ground bus* or to the chassis of the equipment being worked on. Because of this and possible lead poisoning, never hold solder in your mouth. Most production departments now require that *three-wire irons* be used.

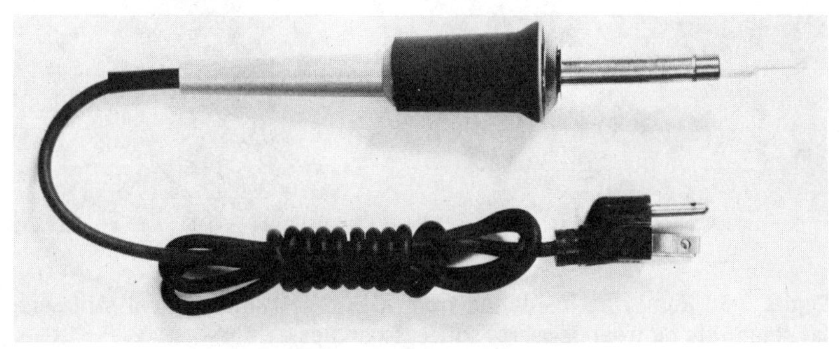

Figure 7-1 Typical soldering iron

Before tinning a wire, remove the insulation with wire strippers (Figure 7-2). To use the mechanical wire stripper, place the wire between the jaws and squeeze the lever (Figure 7-3). To remove the insulation with the thermal strippers, place the insulated wire between the *electrodes* (Figure 7-4) and apply slight pressure to close the electrodes. Rotate the wire slightly while removing insulation to even the cut. *Since some insulations produce a toxic gas when heated, be sure that the room is well ventilated. Military contractors do not approve using any mechanical strippers because mechanical strippers may nick the wire and cause breaks at a later time.*

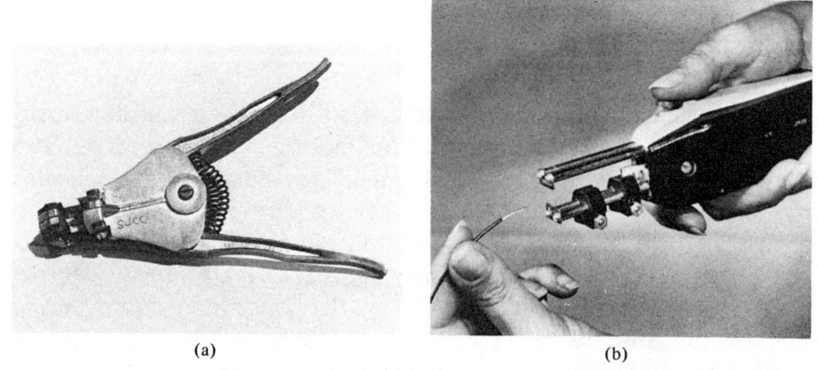

(a) (b)

Figure 7-2 Wire strippers: **(a)** mechanical type, **(b)** thermal type

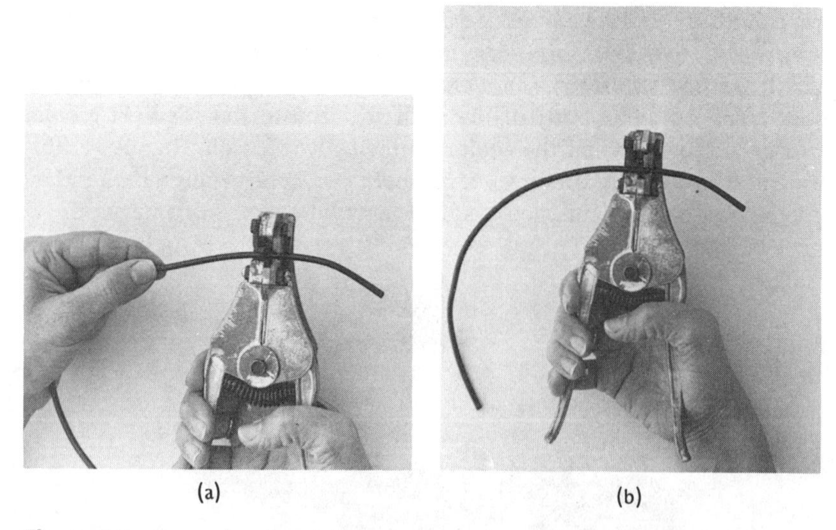

(a) (b)

Figure 7-3 Removing insulation from a wire with mechanical strippers: **(a)** place wire between jaws, **(b)** squeeze handles,

Figure 7-3 (continued)
(c) release handles

(c)

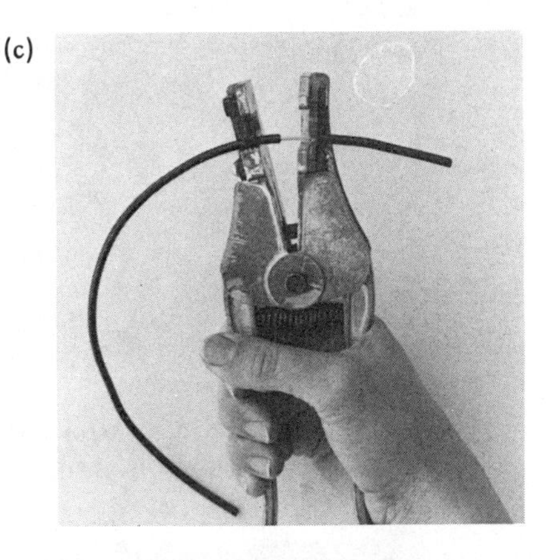

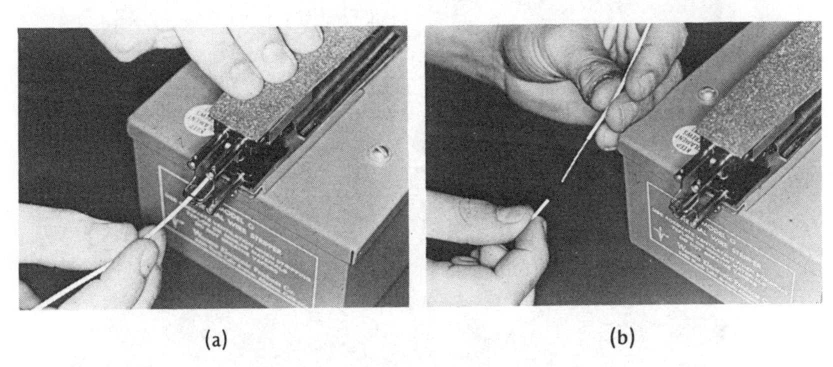

(a) (b)

Figure 7-4 Stripping wire with thermal strippers: **(a)** wire inserted in stripper, **(b)** removing cut by heat insulation

To prepare a soldering iron tip for tinning, it must be heated and cleaned. Clean the tip by dipping it in flux or rubbing it on a flux-soaked sponge. Once the iron has been heated and cleaned, apply a thin film of solder to the tip and wipe away excess solder (Figure 7-5). The film of solder on the iron tip aids conduction of heat to the surfaces being soldered. *Remember,* a soldering iron will not function properly if it has not been tinned as just described.

Figure 7-6 shows the correct method of tinning a wire. The anti-wicking tool supports the wire and prevents solder from running (*wicking*) under the insulation. Each strand of wire must be free of solder at the point where the insulation ends to permit inspection for nicks or cuts in the wire. A heat sink such as shown in Figure 7-7 may be used in place of the antiwicking tool.

To tin the wire, dip the solder into a bottle of liquid flux and smear the flux on the exposed wire. Then melt a drop of solder, called

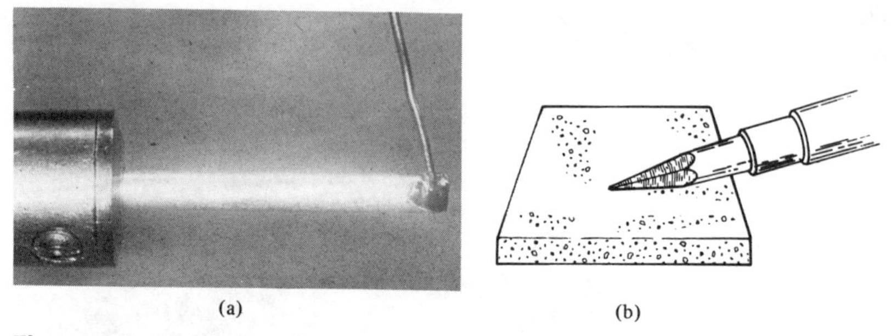

(a) (b)

Figure 7-5 Tinning on iron: **(a)** a film of solder is applied to the iron tip, **(b)** wipe excess solder from iron tip

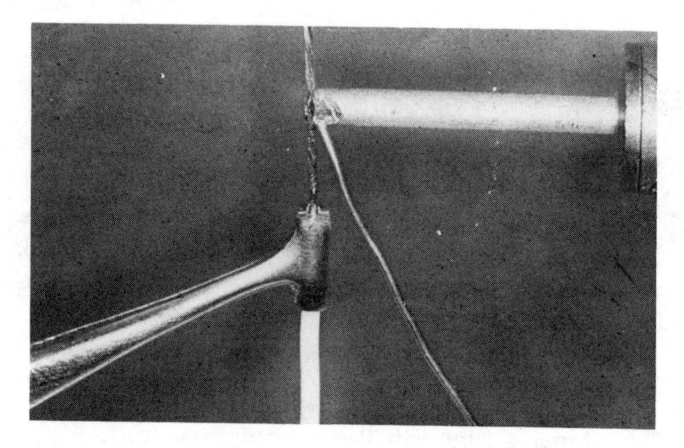

Figure 7-6 Using an antiwicking tool to prevent solder from running under the insulation

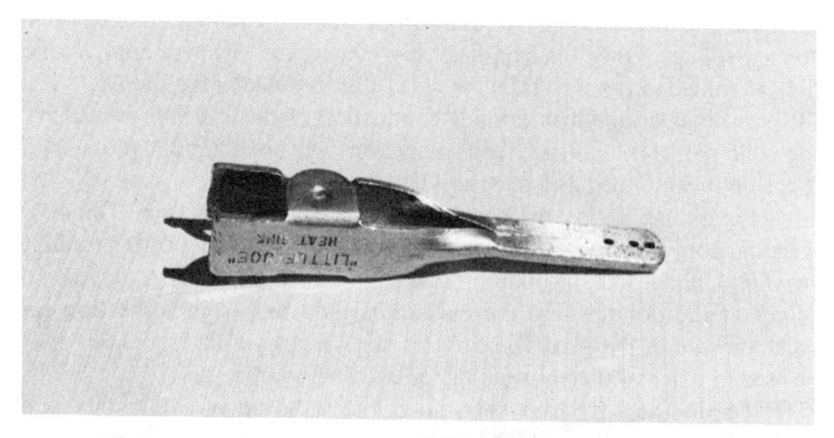

Figure 7-7 A heat sink can be used as an antiwicking tool

a *solder pot,* on the clean tip of the iron (Figure 7-8). Add solder to the iron as you pull the wire slowly through the melted solder (Figure 7-9). Then remove the heat sink and clean the flux from the tinned wire with alcohol on an *acid brush* (Figure 7-10). Finally, examine the wire to see that each strand is visible and that solder did not wick under the insulation.

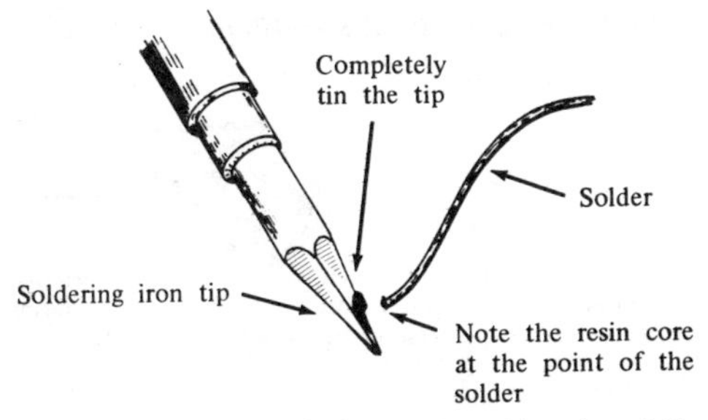

Figure 7-8 A solder pot is formed on the tip of the iron

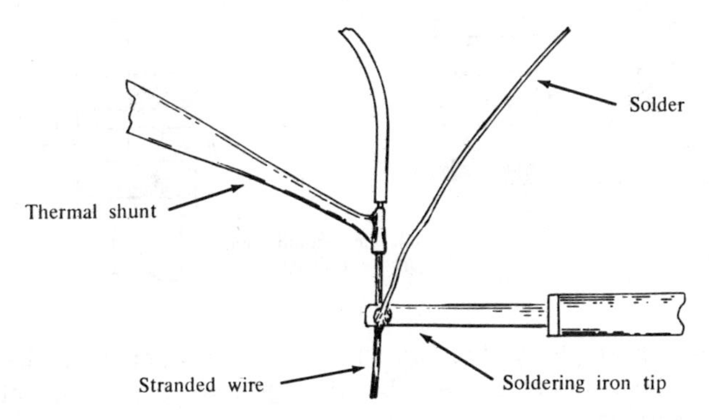

Figure 7-9 Solder is added as the wire is pulled through the solder

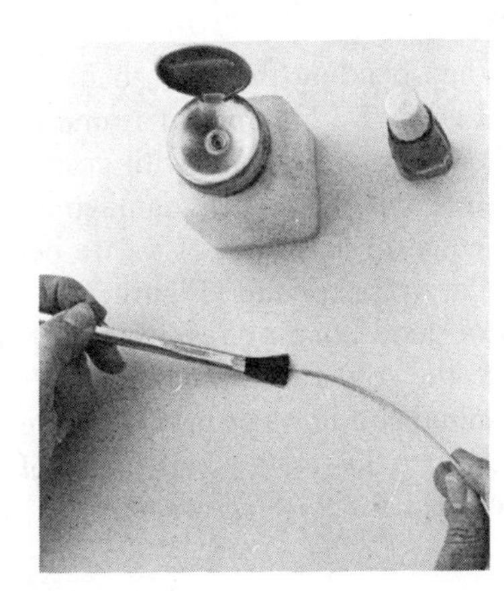

Figure 7-10 Clean flux from the tinned wire with an acid brush

7.3 SOLDERING TO A TERMINAL— GENERAL GUIDELINES

When soldering to a *turret* terminal, wrap the tinned wire 180 to 270° around the terminal, that is, between ½ and ¾ round (Figure 7-11). Wire-bending specifications are usually given in terms of degrees or fractions of a turn on the equipment *specification* sheet.

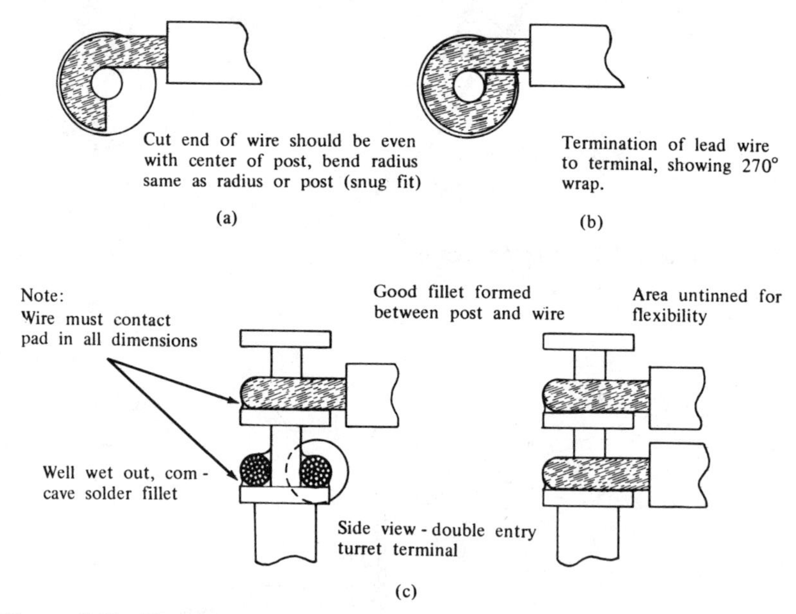

Figure 7-11 Examples of the correct methods of wrapping a wire around a terminal: **(a)** 180 degree wrap, **(b)** 270 degree wrap, **(c)** connecting two wires to a terminal

When bending the wire to fit the terminal, use a rounded instrument (Figure 7-12). Special round-nose pliers are made for this purpose. When handling wire with conventional long-nose pliers use extra care to prevent nicking or damaging the wire. Leave a space approximately equal to the width of a wire between the terminal and the insulation of the wrapped wire (Figure 7-13).

For good bonding results, apply solder to the terminal on the side opposite the soldering iron (Figure 7-14). The solder is melted by the hot terminal and flows around the wire to form a band. *Take care not to use too much solder.* To make sure of this, see that the outline of each conductor in the wire shows through the solder and that the surface is shiny.

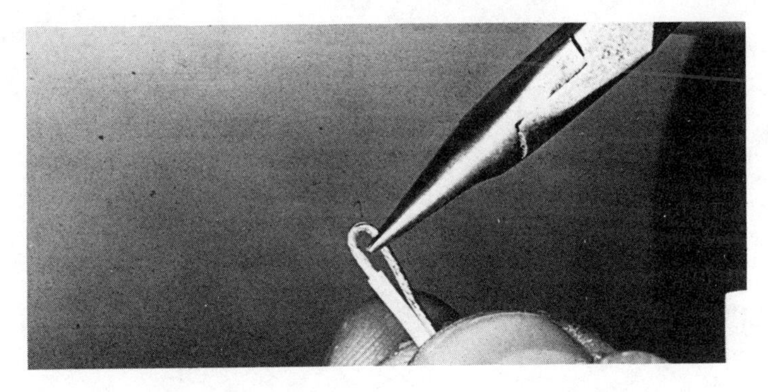

Figure 7-12 Forming a wire to fit a turret terminal

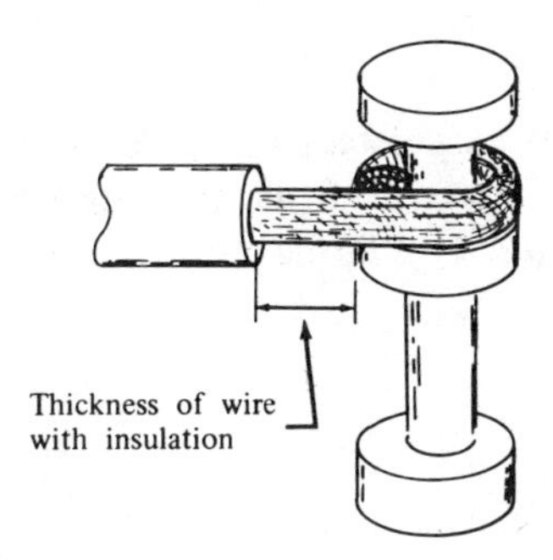

Figure 7-13 Correct insulation dress

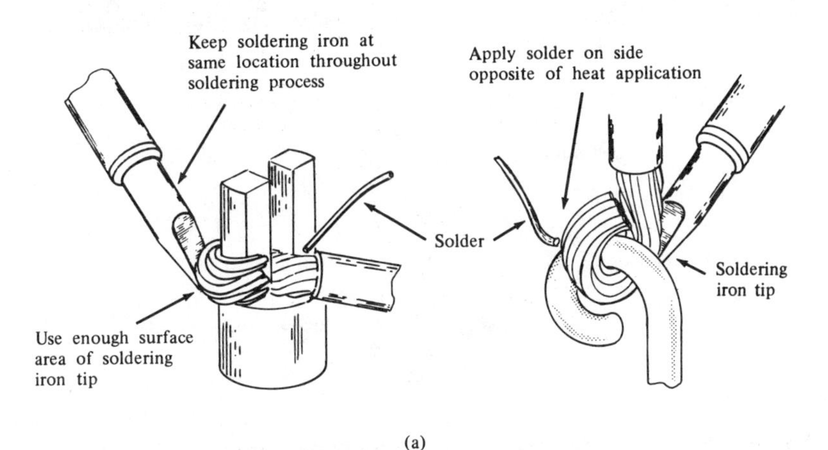

Figure 7-14 Applying solder to the terminal: **(a)** proper applications

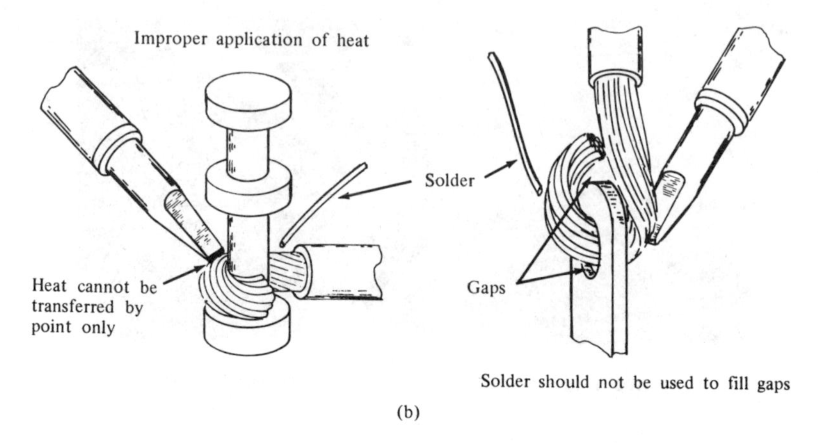

Figure 7-14 (b) improper applications

Figure 7-15 shows unsatisfactory solder connections. Too little solder has been used in Figure 7-15(a), too much solder has been used in Figure 7-15(b), and a *cold joint* has been formed in Figure 7-15(c).

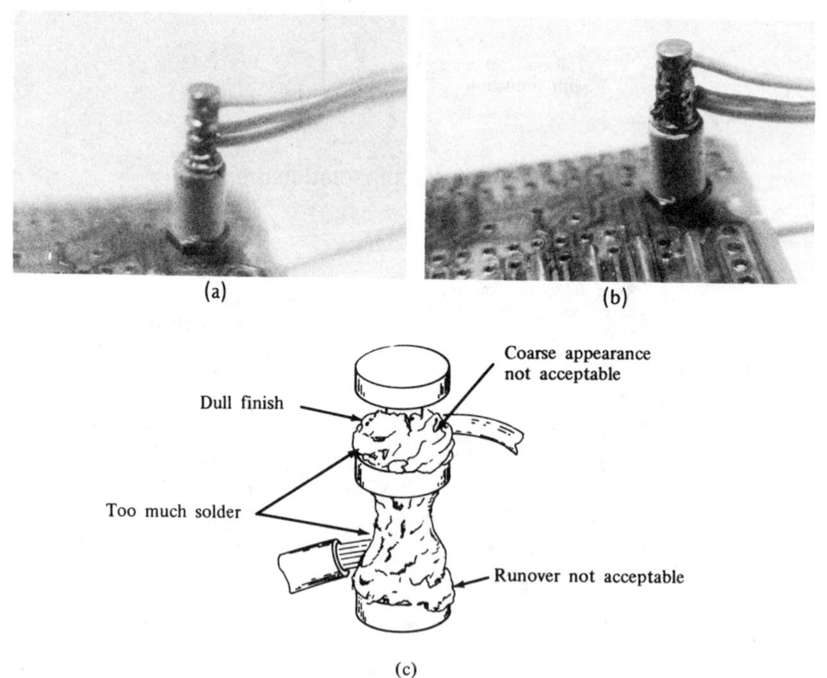

Figure 7-15 Unsatisfactory solder connections: **(a)** insufficient solder, **(b)** excess solder, **(c)** cold solder joint

Figure 7-16 shows a satisfactory connection.

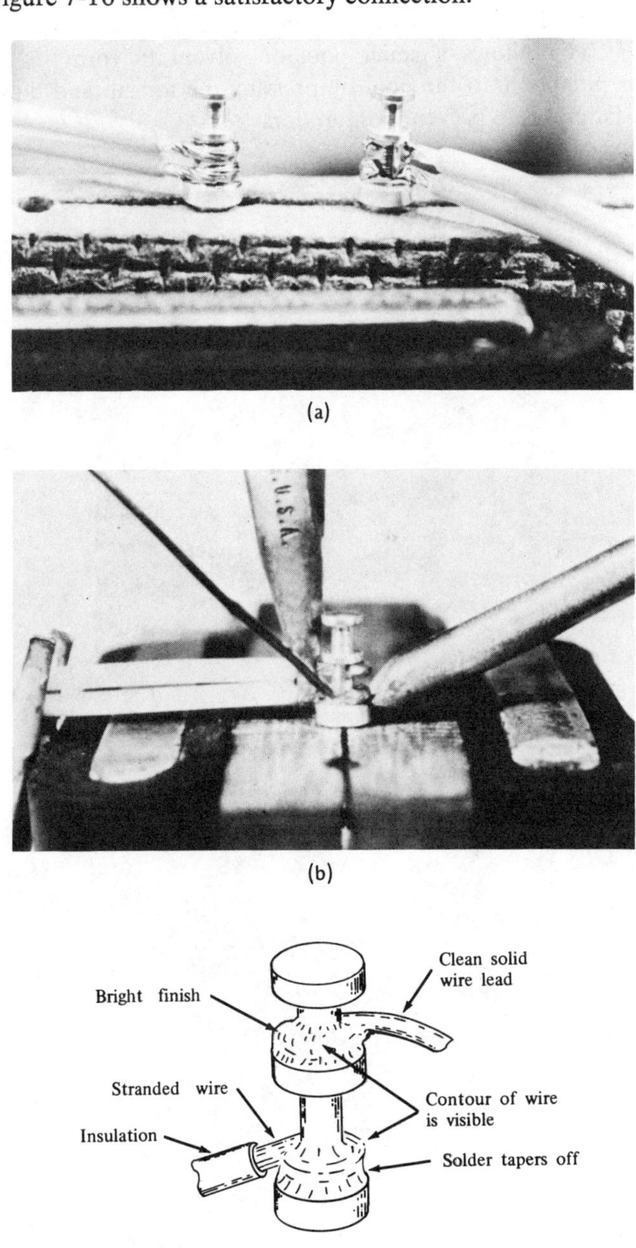

Figure 7-16 Satisfactory solder connections: **(a)** solder just covers wires, **(b)** solder flows into joint, **(c)** details of a good connection

The soldered joint should be cleaned with *ethyl alcohol* or a specified solvent using a brush as shown in Figure 7-17(a). The bottle in Figure 7-17(b) allows a small pool of solvent to form in the top. Solvent is pumped into the pool by pressing the top up and down. The lid of the jar closes to prevent *evaporation.*

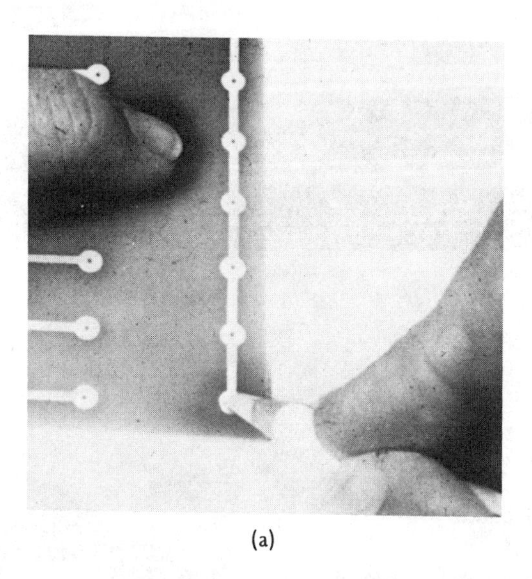

(a)

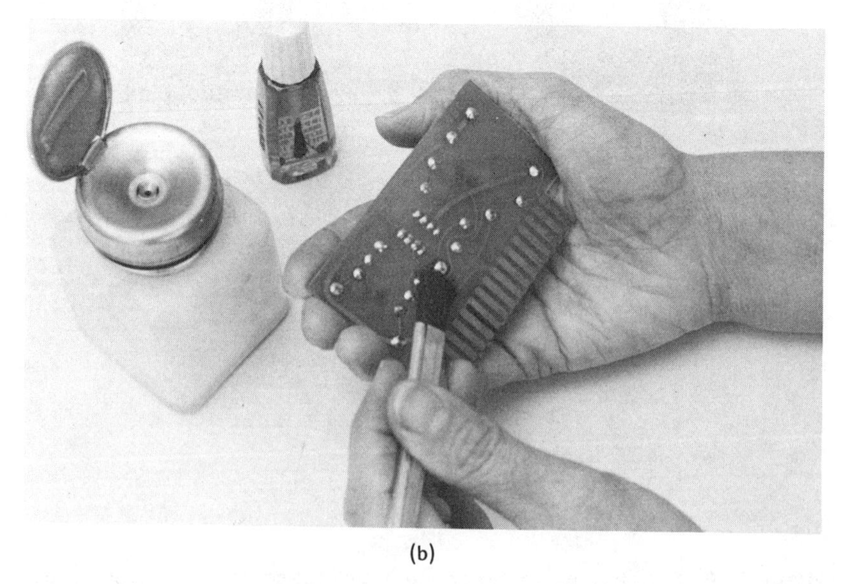

(b)

Figure 7-17 Soldering cleaning tools: **(a)** brush, **(b)** flux cleaner bottle

7.4 SOLDERING BIFURCATED TERMINALS

When soldering *bifurcated terminals,* either insert the wire in the slot or bend it around the terminal as shown in Figure 7-18. Acceptable and unacceptable connections are shown in Figure 7-19. Various wire connections to bifurcated terminals are shown in Figure 7-20. In Figure 7-20 notice that, when two wires enter a bifurcated terminal from the side, each should be bent 90° and wrapped in the opposite direction.

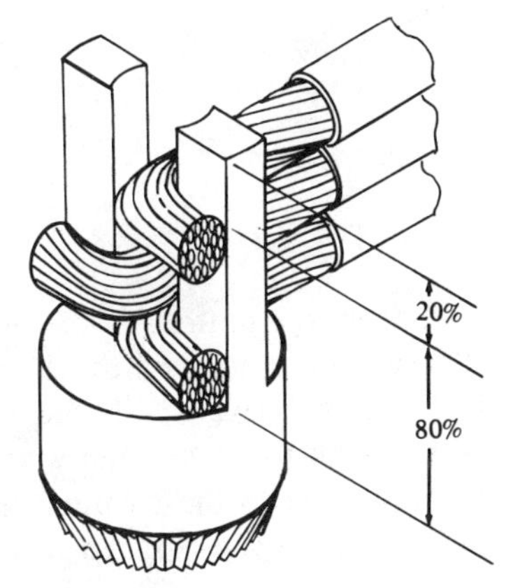

Figure 7-18 Connecting to a bifurcated terminal

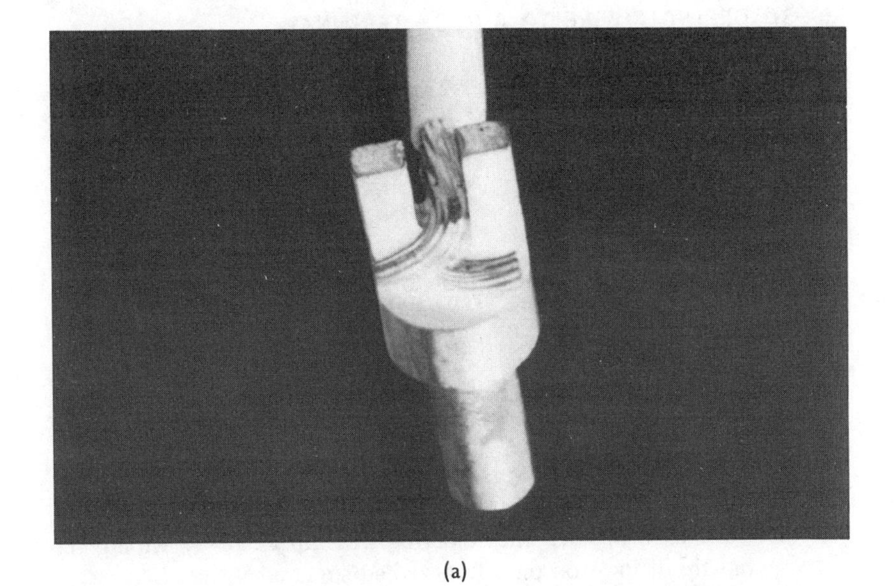

(a)

Figure 7-19 Bifurcated terminal connections: **(a)** acceptable,

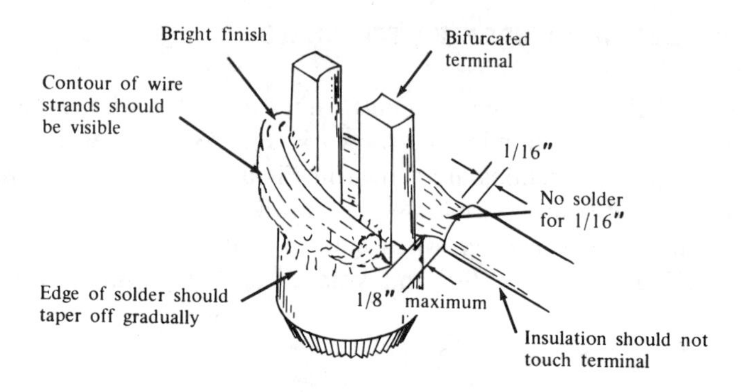

Figure 7-19 (b) unacceptable

After feeding a wire from the bottom of the bifurcated terminal as in Figure 7-20(b), fill the excess area of the hole with a *filler wire* (Figure 7-21). Cut the filler wire even with the top of the terminal. When necessary, use a heat sink to protect the wire. As with any other terminal, apply the soldering iron to the terminal away from the joint. A very small amount of solder may be applied to the iron to help heat transfer to the terminal.

7.5 SOLDERING A WIRE TO A HOOK TERMINAL

Hook terminals, shown in Figure 7-22, are often used on *relays* or other multiterminal devices. Figure 7-23 details the proper *dress* and connection of conductors to hook terminals. The same general guidelines for terminal soldering apply:

1. Clean the terminal.
2. Tin the wire taking care not to allow solder to flow under the insulation.
3. Place a small pool of solder on the tip of the iron for heat transfer.
4. Touch iron to one side of terminal and melt solder into the other side.
5. Do not overheat or oversolder.
6. Take care that the joint does not move before it is cool.
7. Take care to keep the soldering iron tip clean by wiping the hot tip of the iron on a flux-soaked sponge.

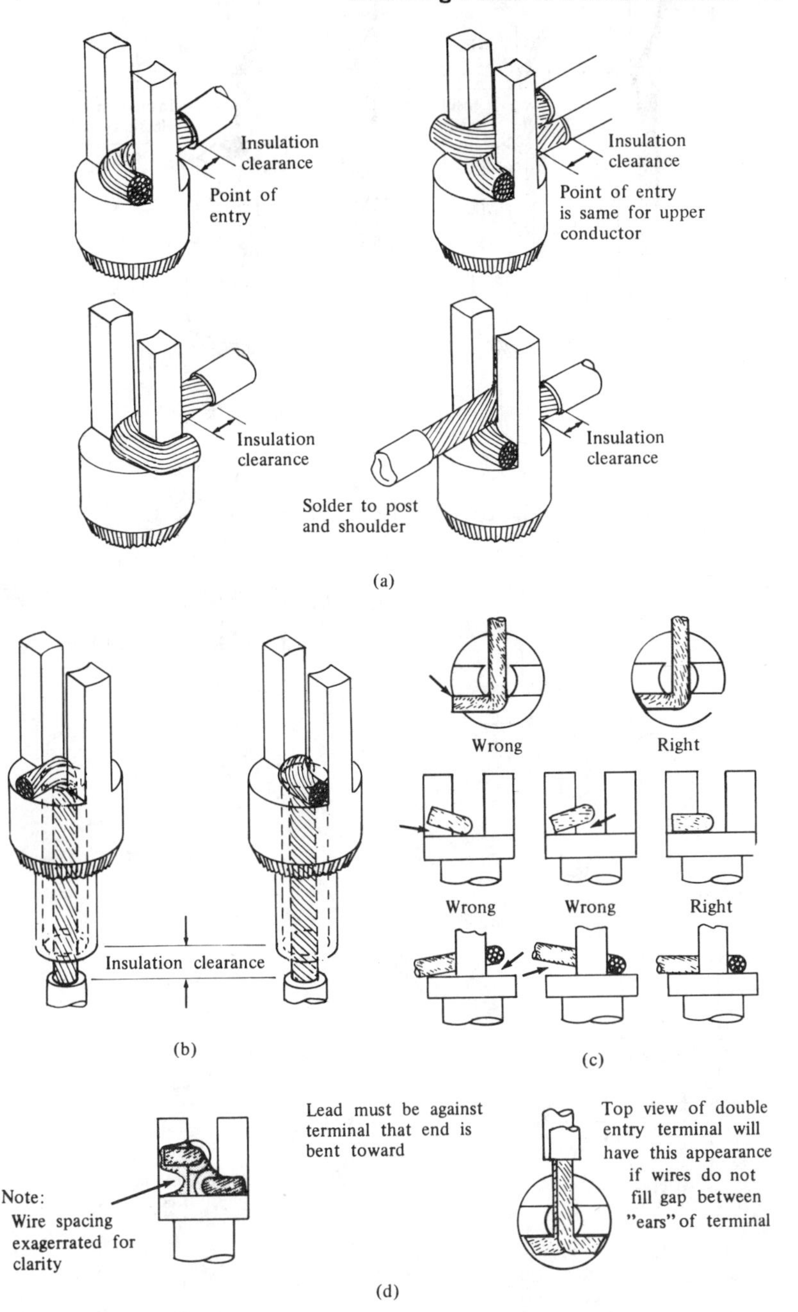

Figure 7-20 Acceptable connections to a bifurcated terminal: **(a)** side route connections, **(b)** bottom route, **(c)** lead position, **(d)** double entry

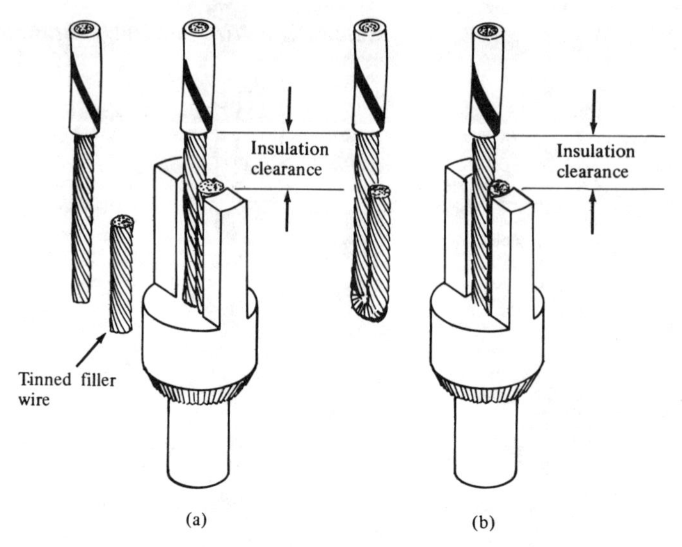

Figure 7-21 Using filler wire

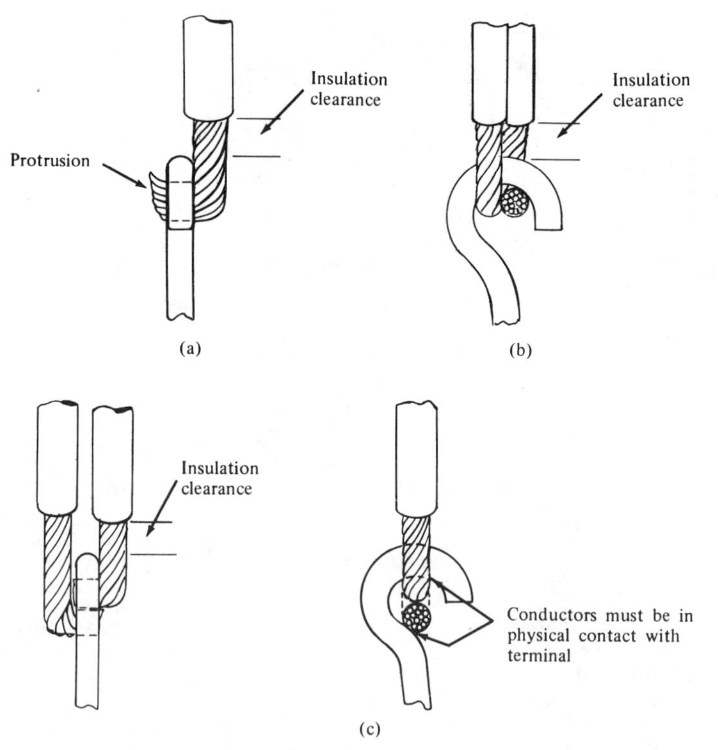

Figure 7-22 Proper connections to hook terminals: **(a)** single-wire conductor, **(b)** multiple-wire conductor, **(c)** less preferred multiple-wire conductor

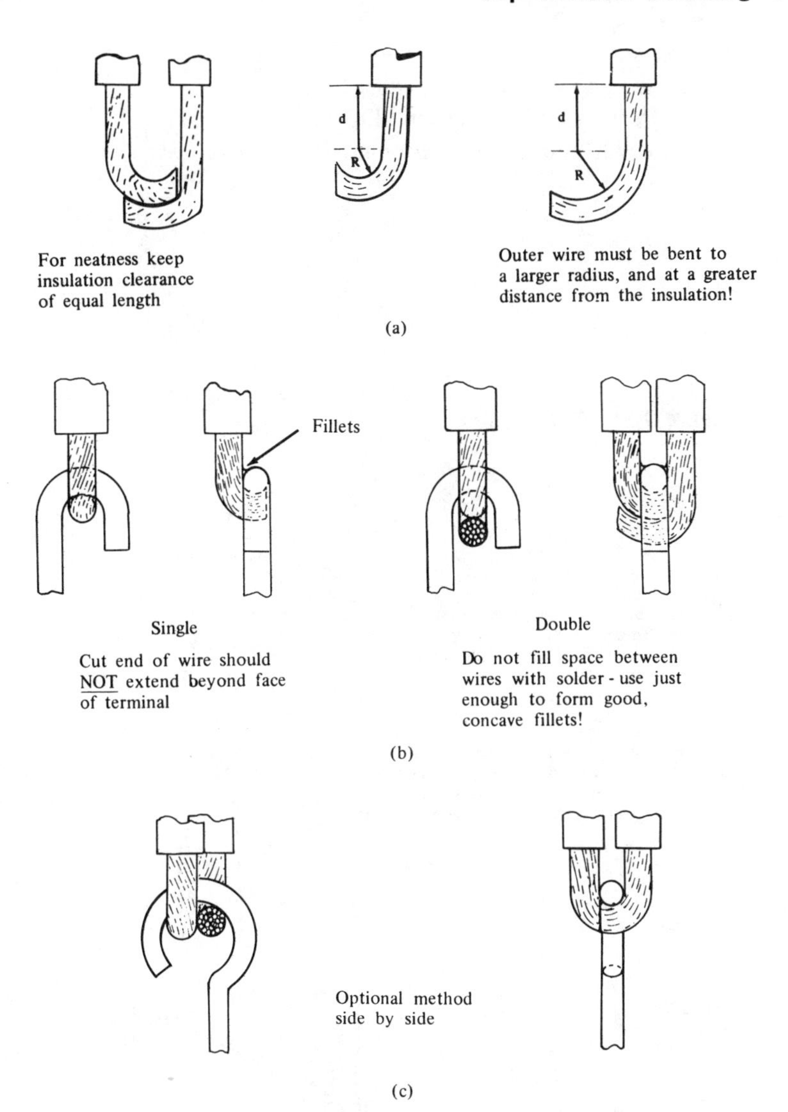

Figure 7-23 Proper wire dress to hook terminals: **(a)** proper bend of wire, **(b)** connection of wire to hook terminals, **(c)** optional method of connection

7.6 CUP TERMINAL SOLDERING

The cup terminal (Figure 7-24) is a hollow tube with a closed bottom. To make a wire connection to a cup terminal:

1. Prepare the wire by stripping and tinning.
2. Fill the cup with solder (Figure 7-25).
3. Insert the wire into the cup (Figure 7-25).
4. Allow the cup to cool before moving the wire.

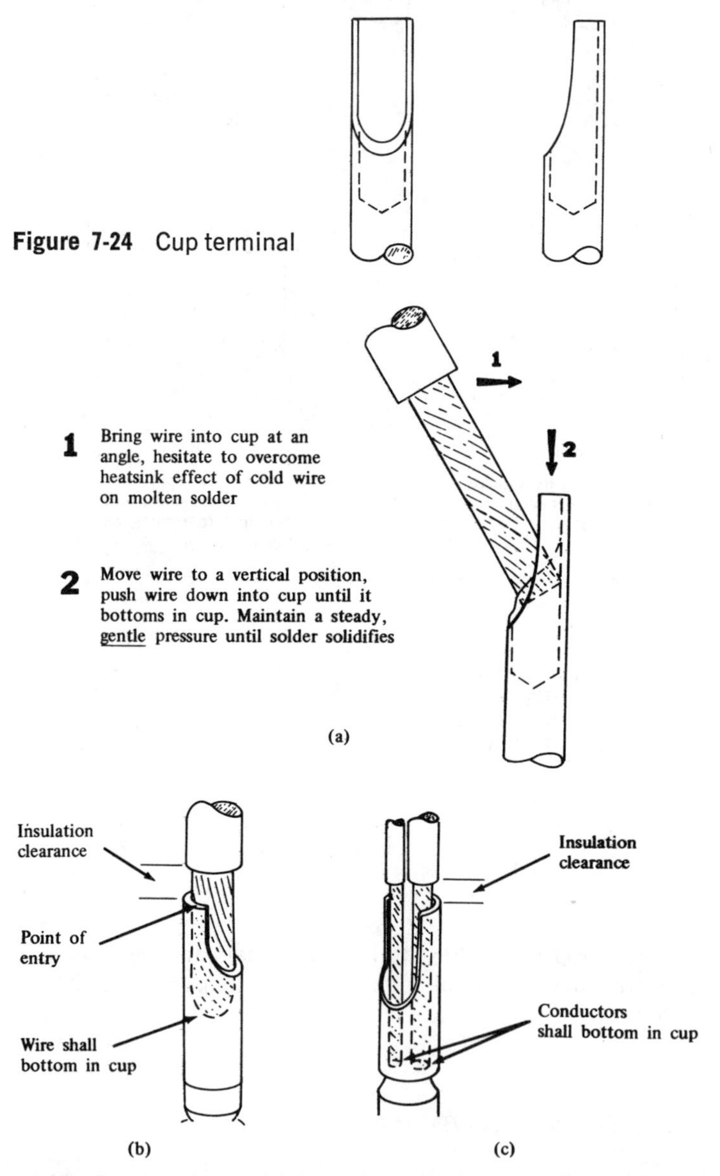

Figure 7-24 Cup terminal

Figure 7-25 Cup terminals: **(a)** inserting wire into cup filled with solder, **(b)** finished details, **(c)** two wires in one cup

The completed connection is shown in Figure 7-25. Connection cups may be soldered with either a soldering iron or *resistance tweezers* (Figure 7-26). In either case, take care to prevent wicking the solder under the insulation.

Figure 7-26 Heating a cup terminal with resistance tweezers

7.7 PRINTED-CIRCUIT BOARD SOLDERING

A typical printed-board is shown in Figure 7-27. Holes for components are shown in different types of boards in Figure 7-28. When specifications require, eyelets are inserted into the holes as shown in Figure 7-29. Figure 7-30 shows several methods of lead termination on the board. The job specification will call out the type of termination required. The same preparation is necessary regardless of the type of termination specified.

1. Clean the board. If you clean the board chemically, handle it with gloves to prevent contamination from the oil on your hands. If you do not clean it chemically, use a rubber eraser to clean the *pad* (Figure 7-31).
2. Prepare the component for insertion into the board using a lead-bending gauge (Figure 7-32).
3. Place the component on the printed-circuit board (Figure 7-33).

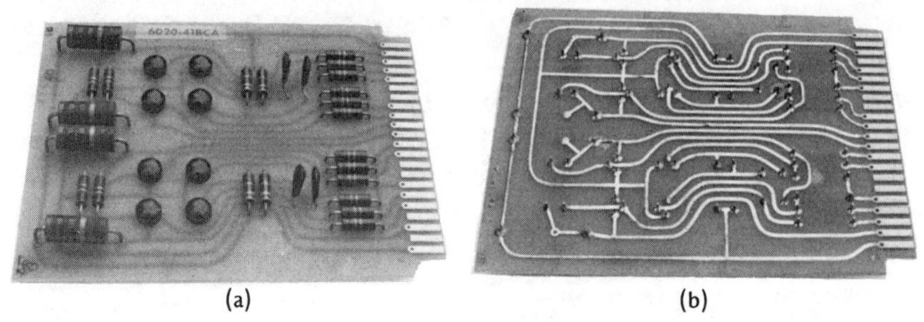

(a) (b)

Figure 7-27 Printed-circuit board: **(a)** typical board, **(b)** detailed view of the board

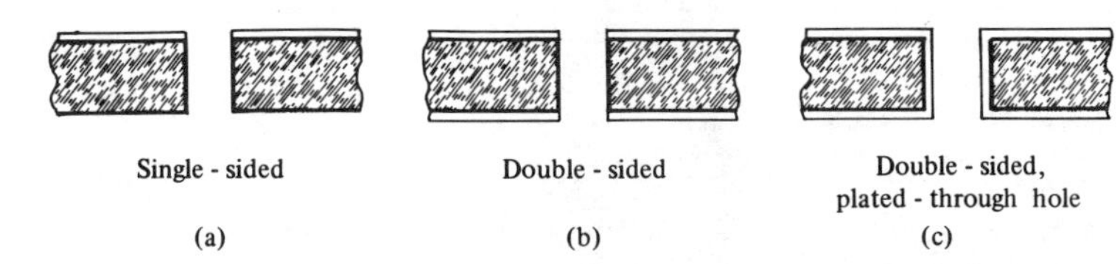

Figure 7-28 Holes in different types of printed-circuit boards

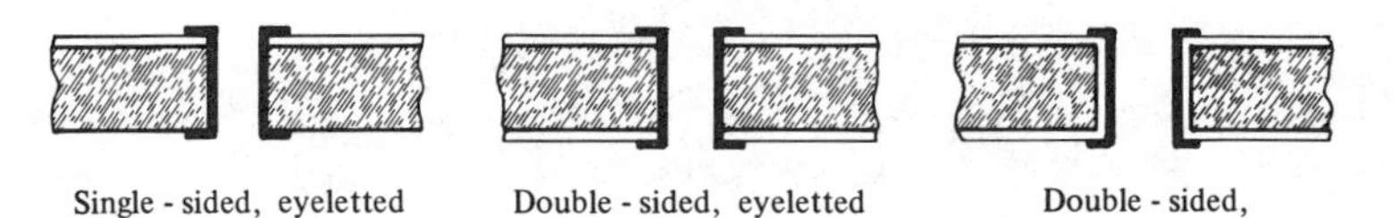

Figure 7-29 Eyelets in different boards

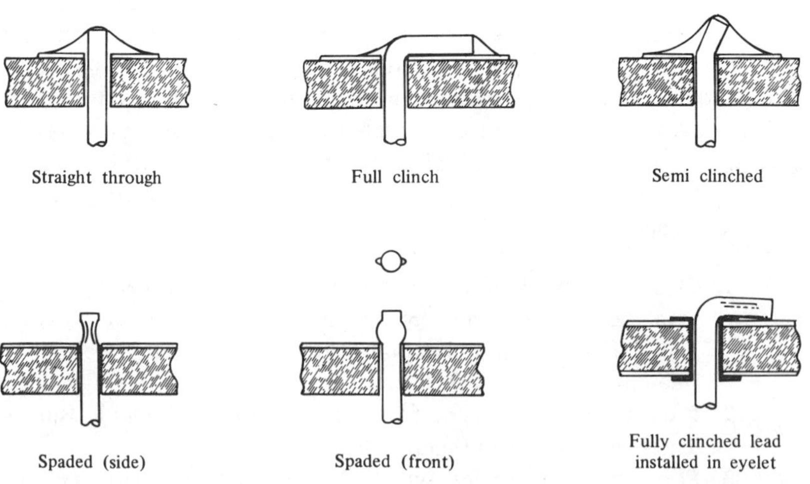

Figure 7-30 Methods of lead termination on a printed-circuit board

4. Heat the junction and apply solder to the side opposite the iron (Figure 7-34).
5. The solder should just cover the lead so that its outline shows (Figure 7-35).
6. The final junction should appear as shown in Figure 7-36. The pad is completely covered with the solder forming a small cone up the edge of the lead [Figure 7-36(b)].
7. Clean the excess rosin from the junction with an approved solvent (Figure 7-37).
8. Inspect each soldering weld for cold soldering, excess solder, or too little solder.

Figure 7-31 Cleaning a terminal pad on a printed-circuit board with an eraser

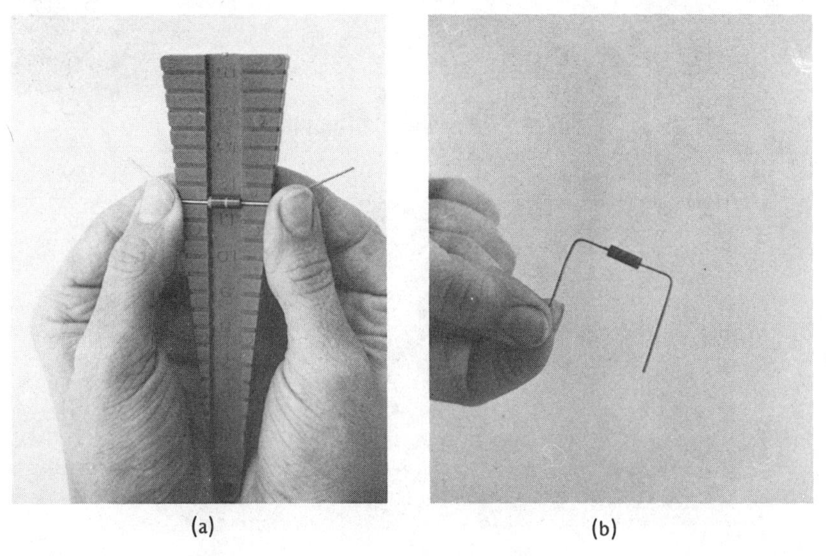

(a) (b)

Figure 7-32 Preparing a component for a printed-circuit board: **(a)** using a lead-bending gauge, **(b)** the prepared component

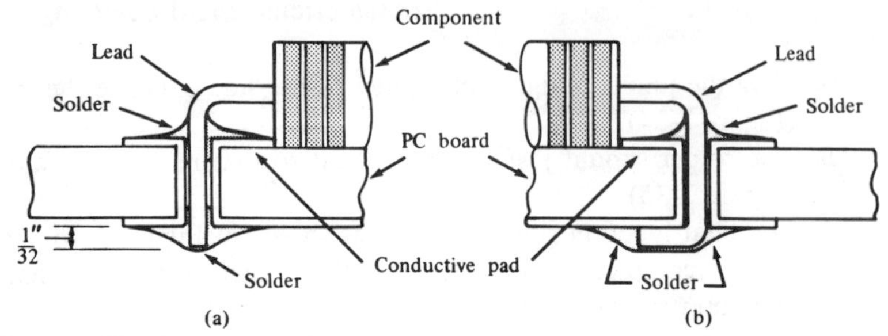

Figure 7-33 Component in place on the printed-circuit board: **(a)** special, **(b)** standard

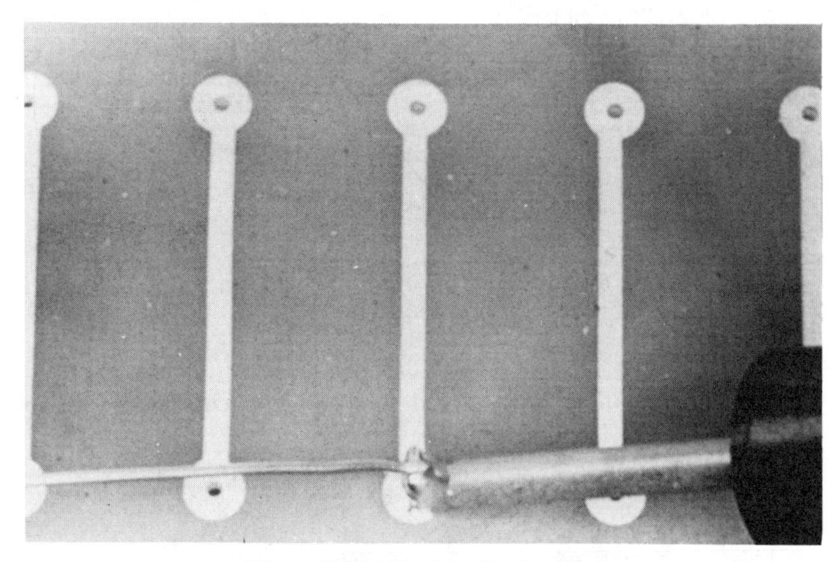

Figure 7-34 Heating the junction

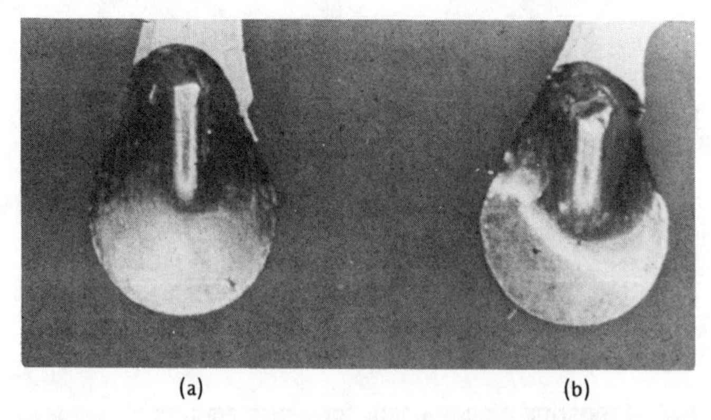

Figure 7-35 Solder connection: **(a)** acceptable, **(b)** unacceptable

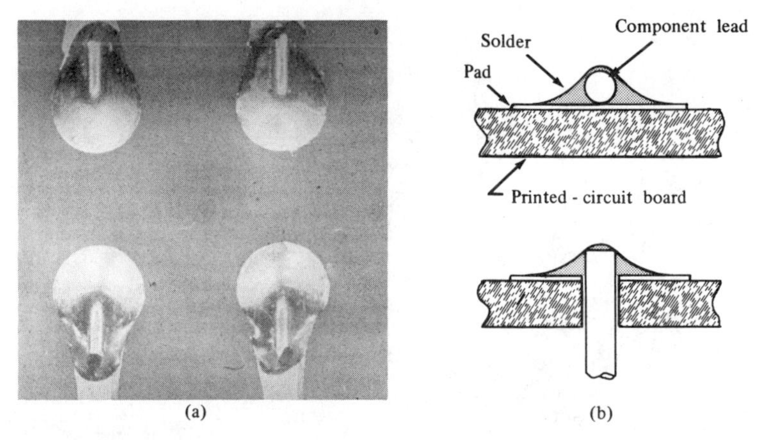

Figure 7-36 Samples of final junctions: **(a)** acceptable solder connection, **(b)** side views of two termination methods

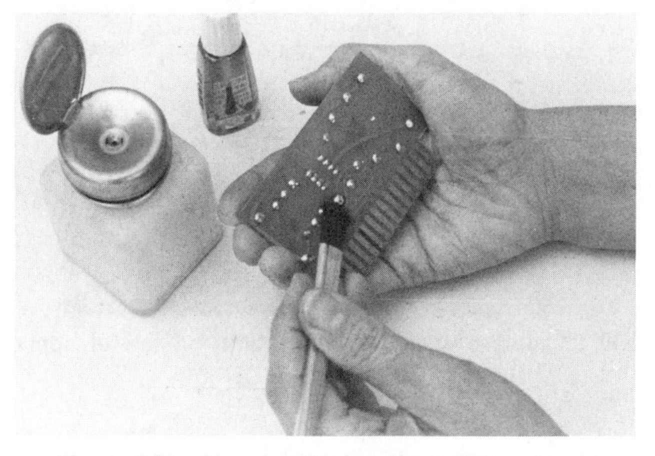

Figure 7-37 Cleaning the junction with a solvent

7.8 PROTECTING HEAT-SENSITIVE DEVICES

Most semiconductor devices can be damaged by excessive heat. To prevent this, a heat sink (a device to take away heat) should be used between the junction to be soldered and the device. A good type of heat sink, shown in Figure 7-38, connects to each side of a diode. A heat sink can also be used as an antiwicking tool (Figure 7-39) to prevent solder from damaging insulation or wicking up the wire under the insulation.

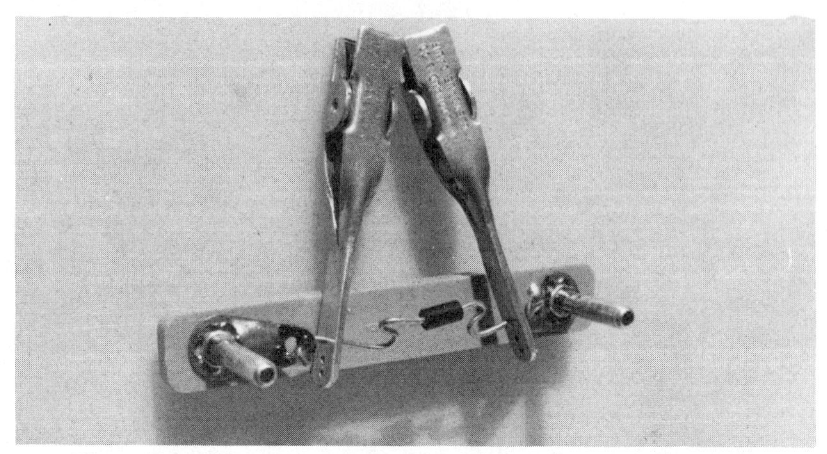

Figure 7-38 Protecting a heat-sensitive device with heat sinks

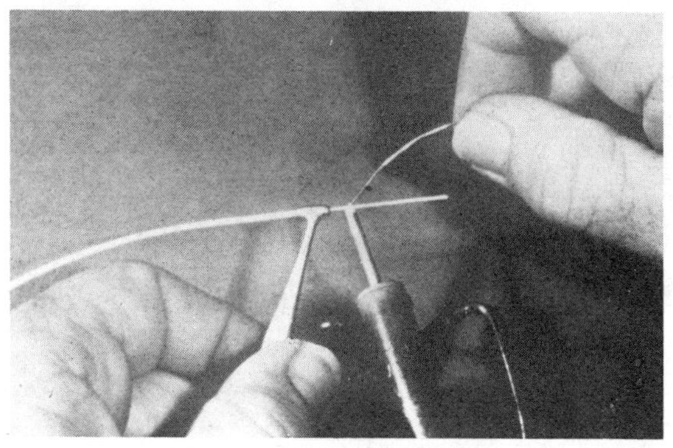

Figure 7-39 Holding a sensitive connection for careful application of heat

7.9 REWORKING AND UNSOLDERING

Repairing electronic equipment by replacing parts requires special care to prevent soldering iron damage to components, wires, or the printed-circuit board. Printed-circuit (PC) assemblies are more likely to be damaged by excessive heat than ordinary chassis assemblies. Therefore, we must learn to handle PC boards carefully in both the soldering and unsoldering procedures. The most important precautions are using the right amount of heat and not using too much solder.

Several methods are used to remove a component from a PC board. One method is to use a *braid* to remove the solder: (1) dip the tip of the metal braid into soldering flux, (2) place the braid over the *pigtail* lead on the part to be removed, and (3) touch the tip of the soldering iron to the braid (Figure 7-40). Never apply the heat directly to the trace. As soon as the solder melts, the braid draws up the excess. Repeat these steps if the area is not totally free of solder. The part can then be removed easily.

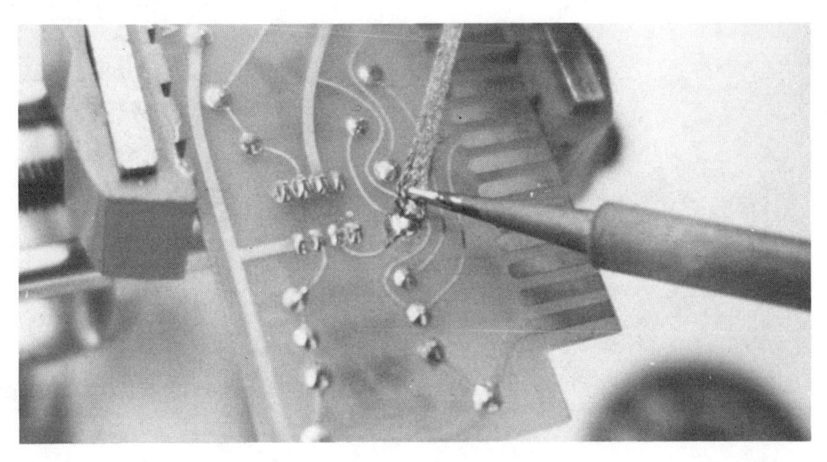

Figure 7-40 Desoldering a component using braid to draw off the solder

A second method of removing solder from a connection is using a suction device. First the solder is heated with a soldering iron, then the melted solder is sucked into the "solder sucker" by air pressure. Figure 7-41 shows three of these devices.

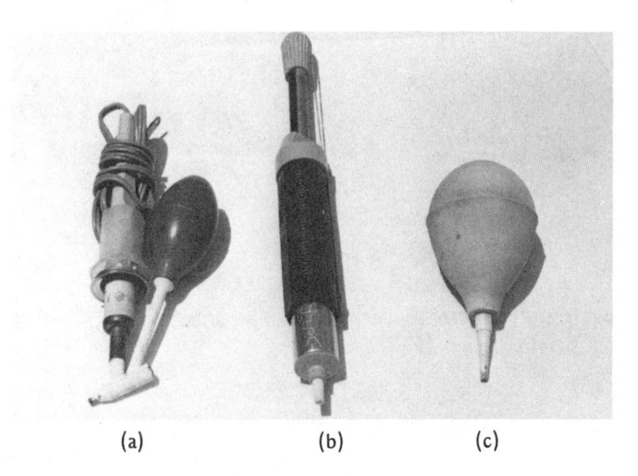

(a) (b) (c)

Figure 7-41 Solder suckers: **(a)** a bulb type on a soldering iron, **(b)** a piston type, **(c)** a bulb type

If you must release two or more terminals to remove the defective component, it may be necessary to heat all the terminals at the same time. This can be done by means of soldering irons that have special tips (Figure 7-42).

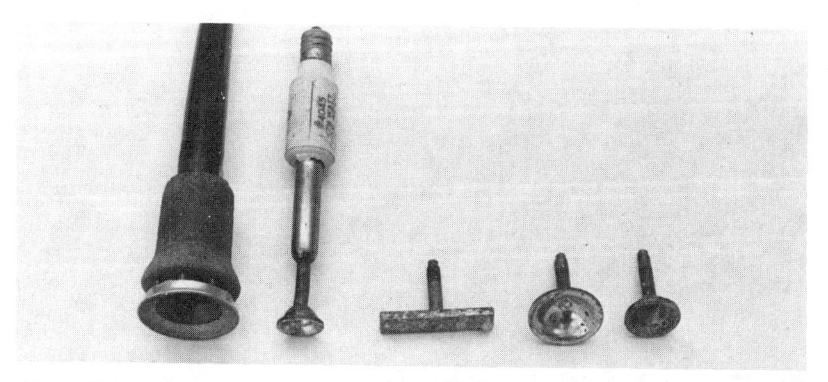

Figure 7-42 Special soldering irons used to heat several terminals at one time

Spring-loaded clips are often used to remove an integrated circuit (IC) while the leads are heated with a special soldering iron tip. Figure 7-43 shows the application of these two devices.

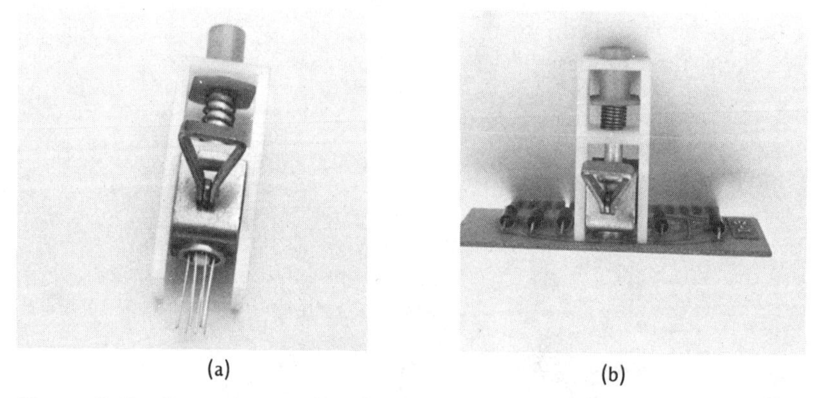

(a) (b)

Figure 7-43 Removing an IC with a special soldering tip and a spring-loaded clip

Sometimes it is difficult to remove a defective component; however, never use force or pressure in an attempt to remove it. If the ends of the leads have been squeezed out of shape, cut them with a pair of diagonal cutters so they may be withdrawn from the board. When the leads are bent, it is usually possible to straighten them before removal. If some solder still remains on the leads, apply the soldering iron for a moment, then withdraw the leads before the solder cools. In any case, it is important to pull the component straight up from the PC board without forcing or twisting it.

Some technicians prefer to leave the cut ends of the component leads in place and to install the new component with a helical soldering aid as shown in Figure 7-44. This soldering aid is a small coil of copperweld wire with an outer coating of solder and flux. It is slipped over the end of a cut lead; the pigtail lead of the new component is inserted into the other end; then the coil is heated with a soldering iron. This creates a solid connection without danger of damage to the PC board.

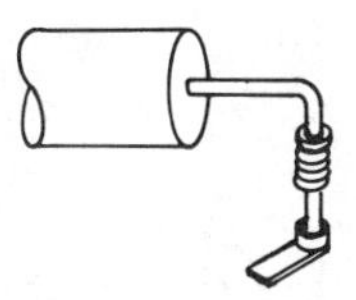

Figure 7-44 Repairing a broken trace with a helical soldering aid

When component leads are being removed from a PC board, *molten* solder may flow into the lead holes. This excess solder must be removed before the new component is installed, otherwise, the end of the lead may tear the pad loose from the board. If you cannot pull out the solder with braid or a solder sucker, use a twist drill of the proper size and carefully drill out the solder from the pad side of the board. Never drill out solder from the component side of the board or the pad is likely to be torn loose.

When a *trace* (printed-circuit conductor) is damaged, it is not always necessary to discard the PC board. For example, it may be practical to remove the loose or charred portions of the trace and replace it with pieces of tinned copper wire bent at both ends. The wire must be long enough to bridge the damaged section of trace. Then drill two holes of correct size near the trace, but not through it, to pass the wire through the PC board. Then insert the wire and bend the ends over the trace. Finally, solder the wire to the trace. The conduction path will now operate as it did before the section of trace was damaged (Figure 7-45).

When repairing equipment, take care not to burn wires and components with the soldering iron and not to let excess solder flow into components or across conductor paths.

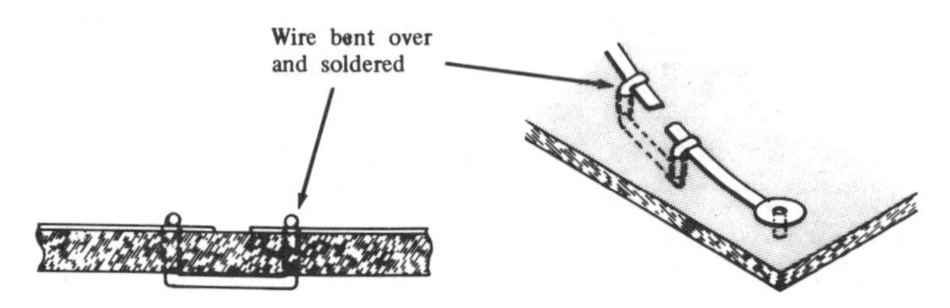

Figure 7-45 Repairing a broken trace with a copper bridge wire

A summary of acceptable soldering connections is shown below. Compare some of your solder work to the pictures.

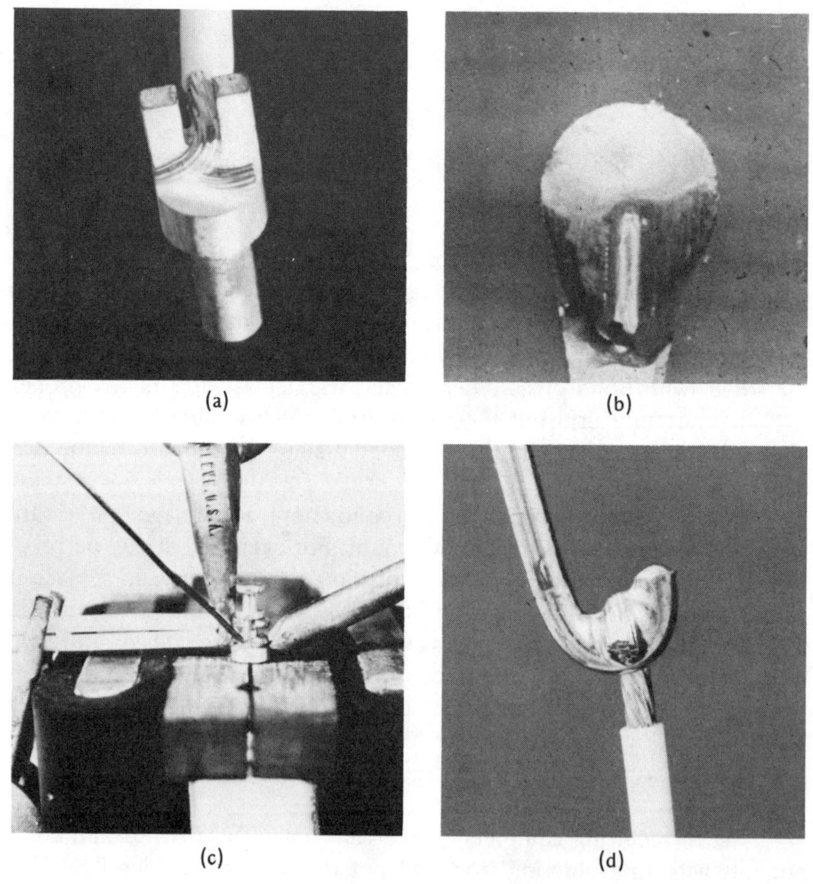

Figure 7-46 Acceptable soldering connections

KT'S KEY TERMS TO REMEMBER

Accelerates Increases in speed or rate of development.

Acid Brush A hair or synthetic bristle brush used to clean the rosin from a completed solder joint or tinned wire.

Alloy A composition of two or more elements. Solder is an alloy of tin and lead.

Bifurcated Terminal A lengthwise, slotted stud used in printed circuits to provide more than one point for connections.

Braid A woven metal tube used as a covering for a conductor. It can be used to create a wicking effect for removing solder.

Cold Joint A solder connection that was not done correctly. When the applied heat is insufficient to cause a metallurgical bond, a cold solder joint results.

Common Ground Bus A heavy strap or bar, often copper, used to make a common connection to several circuits.

Conductivity The ability of a material to transmit heat or current. A good electrical connection has good conductivity.

Cored Solder Solder wire manufactured with a core of rosin so the soldering flux is automatically applied as the solder melts. Also called rosin-core solder.

Dress Refers to the neatness of the insulation, the length of the wire exposed, and the form of the wire, i.e., the overall exactness of a circuit.

Electrodes Conducting elements, they serve to heat objects such as thermal strippers.

Ethyl Alcohol A liquid used for cleaning components as preparation for soldering them.

Evaporation Converting liquid to vapor, which passes away.

Exposed Open or uncovered; when air touches the surface of copper we say it is exposed.

Filler Wire Extra wire inserted into a terminal to plug up the space left by the connecting wire.

Flux A material used to improve the soldering process; it helps clean the joint to be soldered. Rosin is the flux most often used in electronics.

Flux Residue The heat applied to the solder and flux causes the flux to bubble, spread, and solidify—this is the flux residue.

Fuses Binds together by the action of melting.

Isolation Transformer A device designed to provide magnetic coupling between circuits while excluding other sources of coupling such as the house voltage.

Jumper Lead A short wire used to complete a circuit temporarily or to bypass it.

Liquefy To melt; to make solder flow is to liquefy it.

Molecule The smallest particle of a substance that retains the physical properties of the substance and is composed of atoms.

Molten Melted.

Oxide A compound that forms on the surface of an element when it has been exposed to air, for example, the green coating that forms on copper.

Pad Terminal point on a printed-circuit board.

Pigtail A small wire extending from either a component or a connection.

Rosin The most commonly used flux.

Relay An electromechanical device that serves as a switch.

Resistance Tweezers An electrical heating device used to heat terminals for soldering. The two elements of the tweezers carry current that heats the terminal. The advantage of this device is that heat energy is developed only when the tips touch the terminal.

Solidifies Turns solid, hardens; when solder cools it solidifies.

Solder Pot A drop of solder on the soldering iron tip; also, a vat used for dip soldering.

Specification An accompanying paper or document that details information about equipment, materials, or processes to be followed.

Three-wire Irons A system with three conductors one of which is a common ground wire for safety.

Tinning To coat the tip of the soldering iron or wire with a thin layer of solder.

Trace A trace is the ribbon of copper conductor or plating that is on a printed-circuit board.

Turret Small, round post inserted in a PC board as a connecting point for one or more wires.

Wicking The flow of solder up under the insulation, like the action of a wick in a lantern when it draws up oil.

TEST

1. What are the two metals in solder?
2. How is solder applied to bond two wires together?
3. What causes oxide to form on a copper wire?
4. What is the purpose of leaving a spot of solder on the tip of the soldering iron when it is in use?

5. State what you believe are the five most important rules to follow when soldering.
6. What is the purpose of the flux?
7. What is wicking?
8. What is the purpose of using a heat sink when soldering?
9. What is a cold solder joint?
10. Describe a bifurcated terminal.
11. Describe a hook terminal.
12. Describe a cup terminal.
13. What are the special precautions for soldering on printed-circuit boards?
14. What is a lead-bending gauge?
15. How is a component removed from a printed-circuit board?
16. How is an integrated circuit removed from a printed-circuit board?

PROJECT 1: CUP TERMINAL SOLDERING

Materials

1. Ten cup terminals
2. Soldering tweezers or soldering iron
3. Soldering vise
4. Antiwicking tool and solder
5. 24 inches of number 20 wire

Procedure

1. Cut the wires into ten equal lengths.
2. Correctly strip approximately ⅜ in. of insulation from one end of each wire.
3. Correctly tin each wire.
4. Place a cup terminal in the vise and heat the terminal with the soldering tweezers or the soldering iron.
5. Place a small amount of flux in the terminal and heat the terminal at the base with the soldering tweezers or the soldering iron.
6. Fill the terminal approximately three-quarters full of solder by placing the tip of the solder in the cup.
7. Insert the tinned wire into the cup (there should be approximately 1/32 to 1/16 in. of space between the top of the cup and the insulation) and remove the heat from the terminal. It

may be necessary to use an antiwicking tool to prevent the solder from wicking under the insulation.

8. Clean each junction and have your instructor inspect your work.

PROJECT 2: PRINTED-CIRCUIT BOARD SOLDERING

Materials and Equipment

1. Soldering iron
2. Lead-bending gauge
3. Printed-circuit board
4. Rubber eraser
5. Ten inch solid copper wire (approximately #20)
6. Five ½ or 1 Watt resistors

Procedure

1. Cut the wire into five equal lengths and clean each length (if the wire has an enamel coat, use sandpaper to clean the ends).
2. Clean each printed-circuit pad with the eraser.
3. Bend the wires and resistors on the bending gauge to fit the holes in the circuit board as shown in Figure 7-32.
4. Connect the wires to the circuit board as shown in Figure 7-30(a).
5. Connect the resistors to the board as shown in Figure 7-30(c).
6. Make the soldering connections as shown in Figures 33, 34, 35 and 36. Be careful not to use excessive solder.
7. Clean each connection with solvent as shown in Figure 7-37 and turn in the completed job to your instructor.

PROJECT 3: DESOLDERING

Materials and Equipment

1. Soldering braid
2. Solder sucker
3. Soldering flux
4. Soldering iron
5. The connections made in Project 2

Procedure

1. Dip the tip of the soldering braid into the flux and place the braid (See Figure 7-40) on the connection to be desoldered. Place the hot tip of the soldering iron on top of the braid and apply a slight pressure. The heat from the braid will melt the solder on the connection which will then flow toward the hot iron. Remove the braid to check that most of the solder has been removed. If solder remains on the connection, clip the end of the braid and repeat the operation.
2. Desolder the remainder of the connections using the solder sucker as shown in Figure 7-41.
3. Have your instructor evaluate your work.

PROJECT 4: SOLDERING HEAT-SENSITIVE DEVICES

Materials and Equipment

1. Soldering iron
2. Solder
3. Two heat sinks or long-nose pliers
4. Six soldering terminals
5. Three diodes or substitutes
6. Soldering flux
7. Solvent
8. Solvent brush

Procedure

1. Connect each diode to a terminal and connect the heat sinks as shown in Figures 7-38 or 7-39.
2. Solder and clean each terminal.
3. Have your instructor evaluate your work.

8: Harnessing

8.0 INTRODUCTION

Electronic components are often connected by several wires. For economy of space, neatness, and ease of identification, the wires are grouped together and tied. Each group of wires is called a "wire harness." A wire harness may be developed in a chassis or *prefabricated* on a *jig board*. A wire harness differs from a cable in that a harness provides for *service loops* of individual wires throughout its length, whereas all the wires in a cable are the same length. A harness is shown in Figure 8-1(a). Remember to learn the KT's at the end of the chapter.

8.1 OVERALL HARNESS CONSTRUCTION TECHNIQUES

When a single unit is being constructed, a harness may be assembled after the conductors (wires) have been located in place and soldered. Often, however, many units requiring the same harness will be constructed. In this case, the first harness is not soldered into its permanent place but is completed, tied, and then removed to be used as a *template* to make other harnesses (Figure 8-1 b). These production harnesses are made on a harness board.

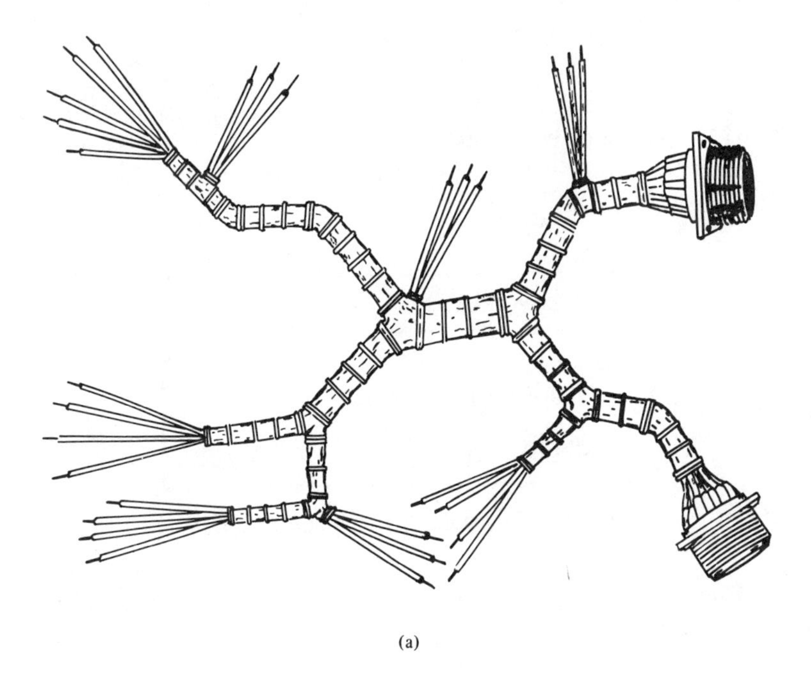

(a)

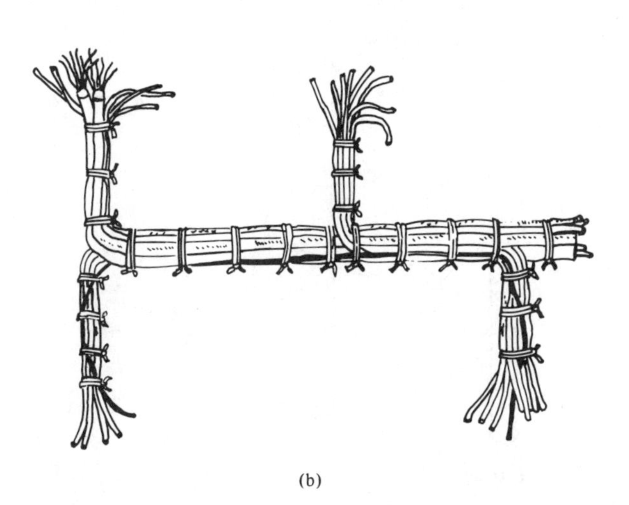

(b)

Figure 8-1 Wiring harnesses: **(a)** wiring harness with terminations, **(b)** harness removed to be used as a template for a harness board

8.2 LACING THE CABLE

Lacing material, which may be nylon or any other similar material, is referred to as cord or twine and is often *impregnated* with wax to prevent slippage. Lacing twine is available in varying widths for various cable sizes as shown in Figure 8-2.

Cable diameter	Cord size
Up to 0.38 in. (9 mm)	# 4
0.44 to 0.75 in. (18 mm)	# 6
0.81 to 1.00 in. (25 mm)	# 9
Over 1.00 in. (25 mm)	# 12

Figure 8-2 Cable-lacing specifications

There are two lacing techniques employed in harness making: (1) the running-lace method, composed of a series of connected loops (Figure 8-3), and (2) the spot-tying method, shown in Figure 8-4. In a properly laced cable, whether the running lace or lace-tying method is used, the ties should be spaced about ½ in. apart or equal to twice the cable diameter if it is larger than ½ in. Special tying techniques are required at breakout points on a cable. Figure 8-5 shows two methods that may be used. The running lace method in Figure 8-3 is being replaced by the spot-tie method (Figure 8-4) and the mechanical spot-tie method introduced later on page 106. These two methods are easier to rework when a cable must be changed.

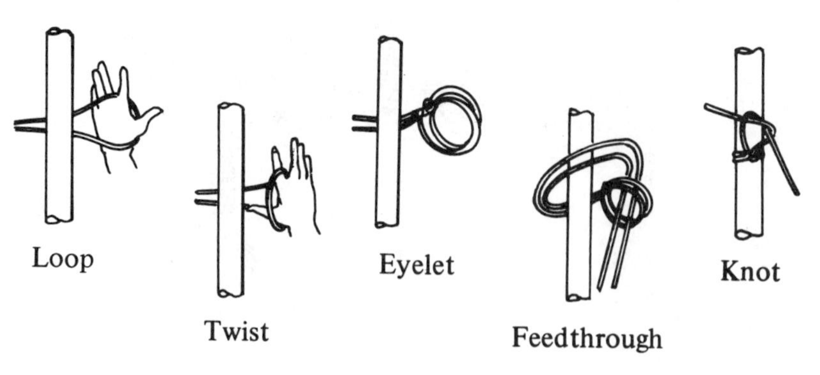

(a)

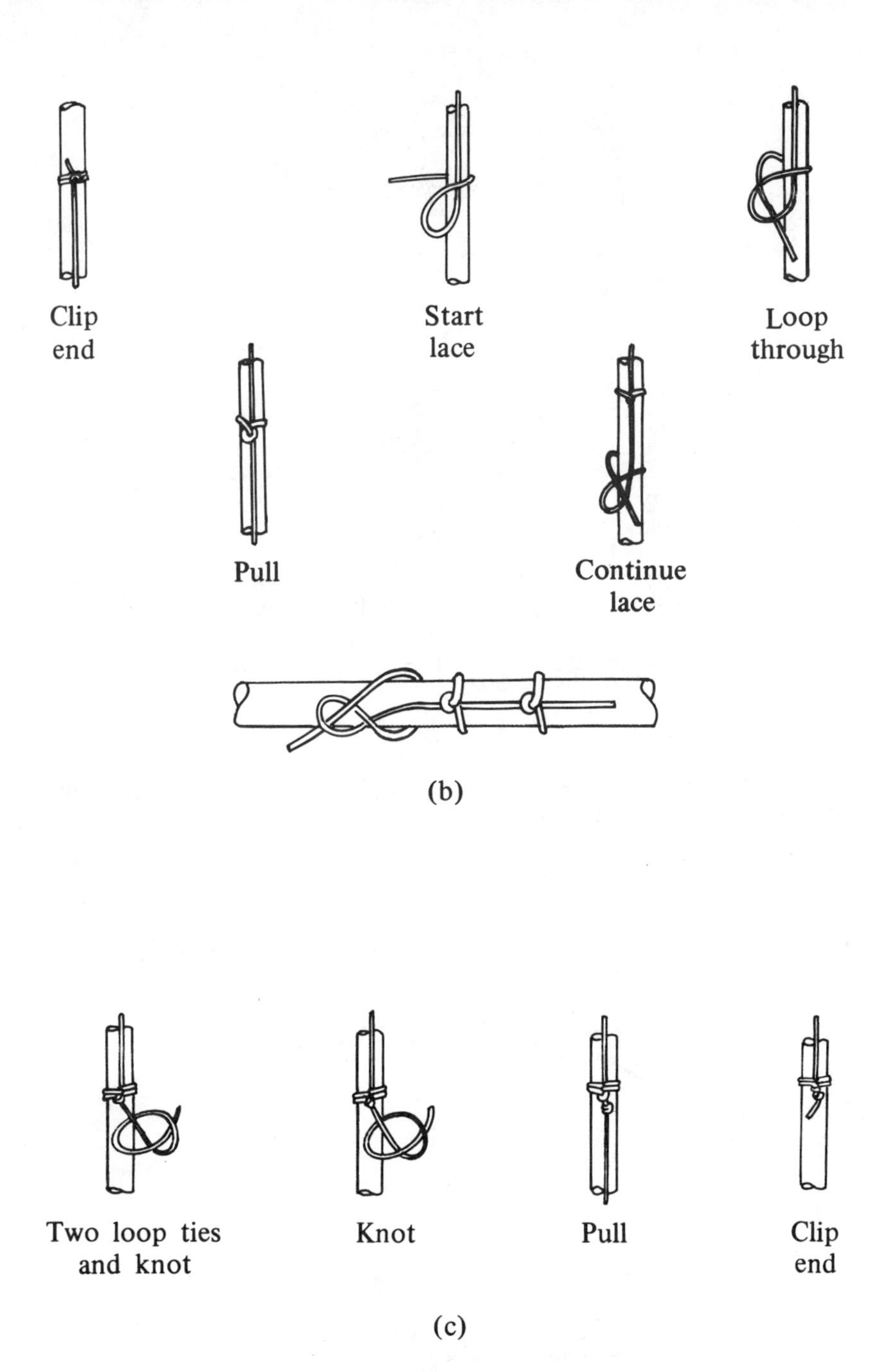

Figure 8-3 Running type of lacing: **(a)** starting tie, **(b)** the running lace, **(c)** the final tie. This type of lace is being dropped due to the time necessary to perform the initial lace and the rework.

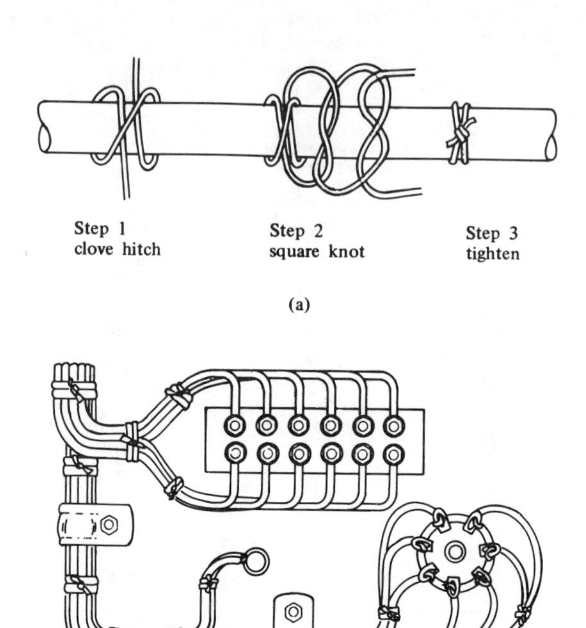

Figure 8-4 Spot-tying type of cable lacing: **(a)** making the tie, **(b)** sample of a cable run with spot tying

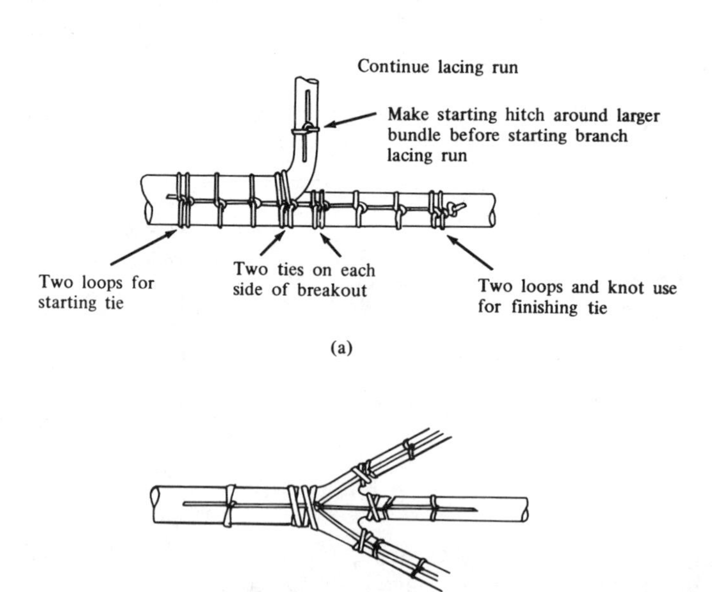

Figure 8-5 Two methods of cable tying for breakout points: **(a)** breakout at a single branch, **(b)** breakout at a three-legged branch

Developing a harness board requires a series of steps that must be performed in a specific order. If the harness is to be duplicated many times, the following sequence is the proper one to use.

1. Construct a mock-up of the equipment if a real sample is not available (Figure 8-6).

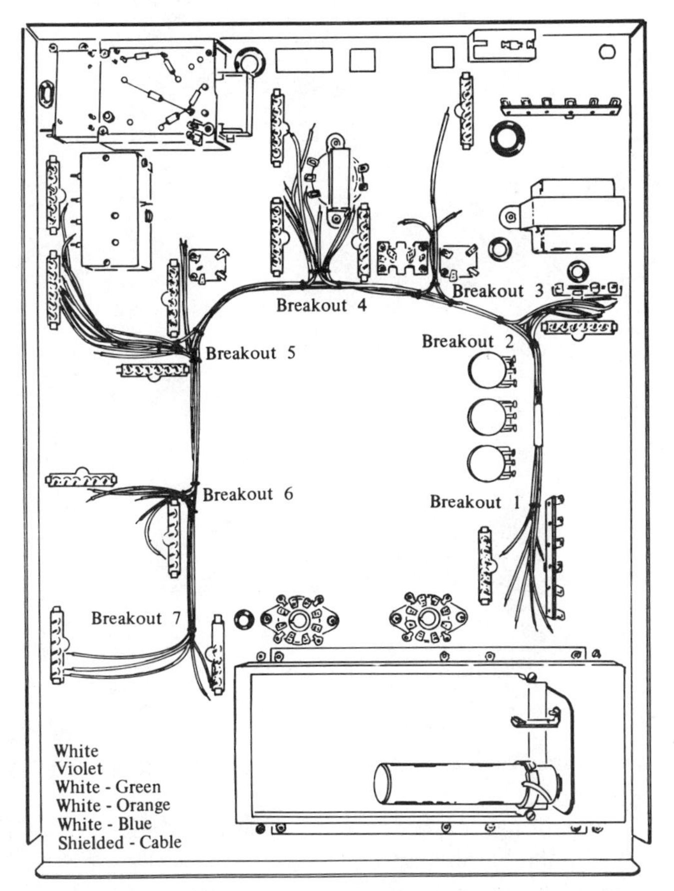

Figure 8-6 Mock-up of equipment for a harness board *(Courtesy of Heath Company)*

2. Attach all connectors to their appropriate contact points. A wraparound hookup may serve since all that is needed is the correct connector length.
3. Lace the connectors into a harness, making sure all cable branches have properly laced breakout ties. Recheck connector lengths.

4. Remove the laced harness, which should look something like Figure 8-7.

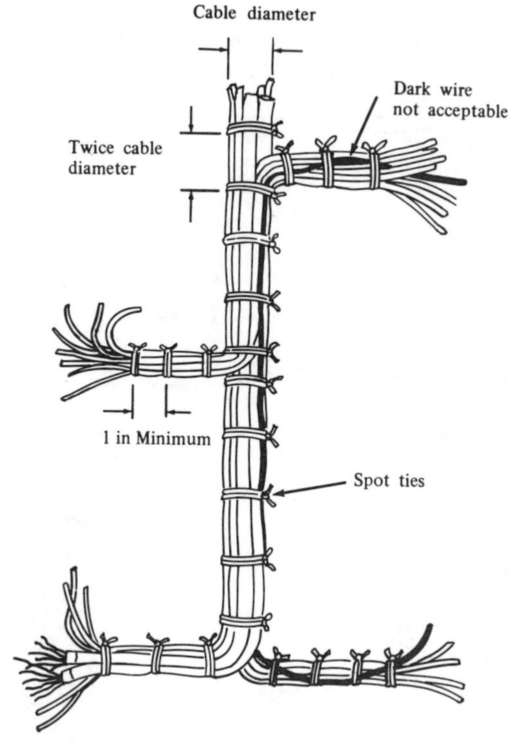

Figure 8-7 A typical wire harness

5. Then make a harness drawing and cement it to a board (Figure 8-8).
6. Attach springs or pegs to the board to serve as guides for forming the harness into the desired shape (Figure 8-9).
7. After placing one set of wires on the board, it becomes obvious that following a specific sequence is preferred so a wire sequence list is then developed. An example is shown in Figure 8-10.

8.3 SLEEVING

Sleeving is a means of holding wires together without lacing. For example, a bundle of wires can be formed into a cable by placing them

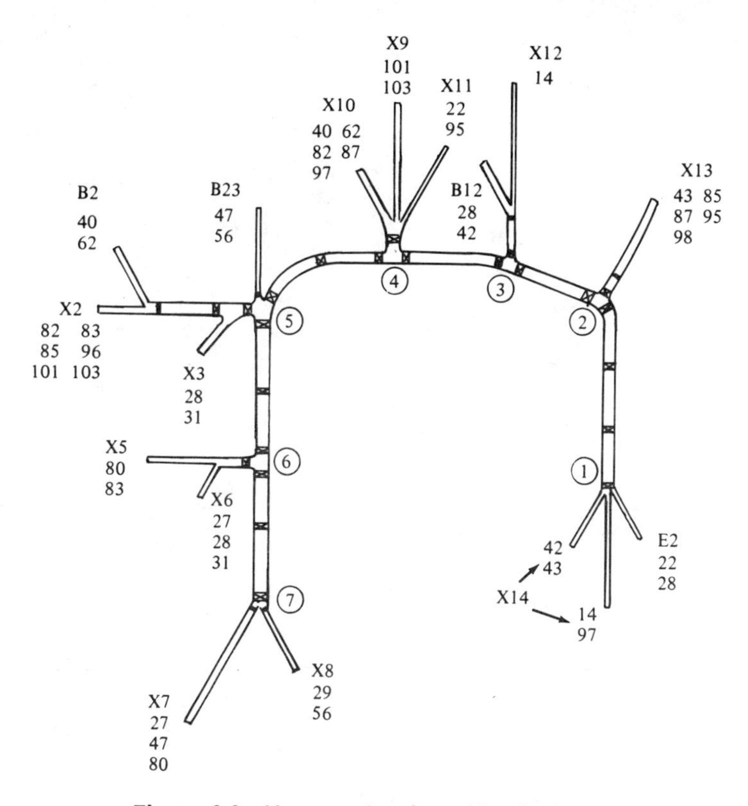

Figure 8-8 Harness drawing with wire layout

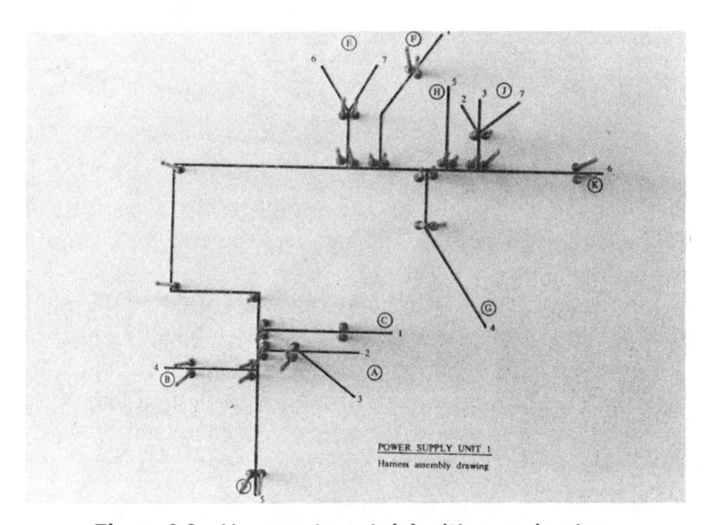

Figure 8-9 Harness board: **(a)** with pegs in place

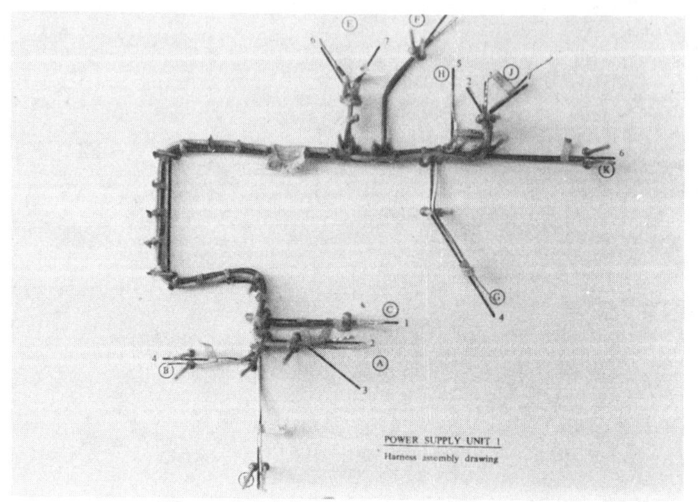

Figure 8-9 (b) with harness laid out

Wire sequence list												Assembly number H - 522-000-001	
												Method plan No.	Id. No. 0
No.		Wire No.	AWG	Color	Lg.	From 0	Wire routing		From (A)	Shielding A	Shielding B	To (B)	Notes
1		62	20	999	12	1		3					
2		72	14	999	20	1		4					
3		82	20	999	7	1		2					
4		85	20	999	10	1		6					
5		50	22	999	14	2		4					
6													

Figure 8-10 A wire sequence list

in a sleeve (Figure 8-11). Sleeves are also good coverings for *splices* or leads from components because they prevent the bare wire from making contact with other wires or parts.

There are three types of tubing commonly used as sleeving: heat-shrinkable tubing, loose tubing, and zipper tubing. Heat-shrinkable tubing (Figure 8-12) is commonly used as insulation for splices. Its installation is simple: (1) cut a piece of tubing the desired length, (2) slip the piece of tubing over one of the wires, (3) complete the wire splice, (4) finally, slide the tube over the splice and heat. The shrinkable tube will seal the bare splice.

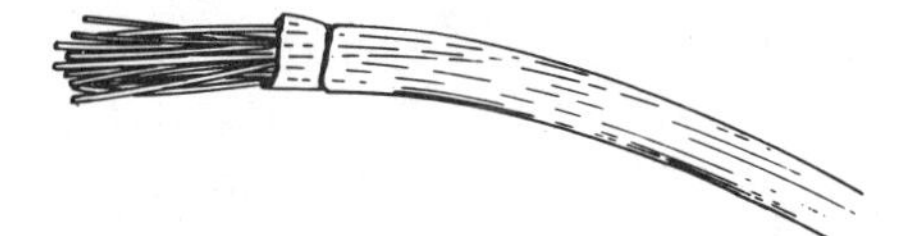

Figure 8-11 Making cable by placing a bundle of wires in a sleeve

Figure 8-12 Applying heat to heat shrinkable tubing with a heat gun

Loose tubing which is used as sleeving over connections, splices, and component wires, is simple to install (Figure 8-13). The following steps, performed in the order given, make the task easy.

1. Determine the length of tubing to be installed.
2. Tie the end of the bundle of wires with lacing twine.
3. Feed the lacing twine through the tubing.
4. Pull the bundle through the tubing. To avoid excessive friction, apply talcum powder to the bundle of wires. This will make the bundle slide through the tubing easily.

Figure 8-13 Loose tubing used to form a group of wires into a cable

Zipper tubing (Figure 8-14) is often used because it is much easier to install than loose tubing. The zipper tubing is placed around the wire and then zipped. This type of tubing usually fits loosely around the cable and is not available in as many sizes as the loose tubing.

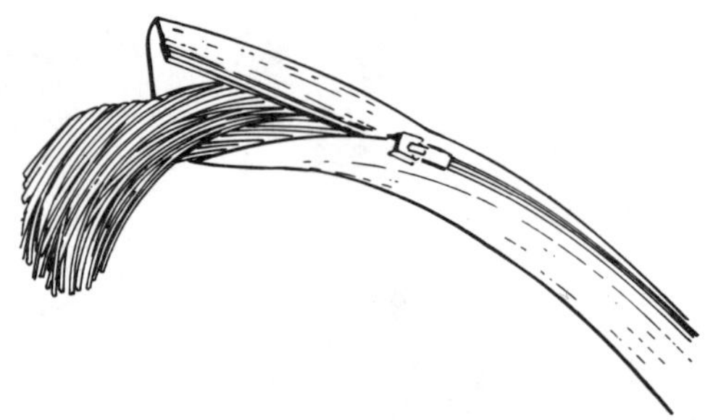

Figure 8-14 Zipper tubing used to form a group of wires into a cable

8.4 STRAPS AND CLIPS

Straps, commonly made of nylon or *plastic,* are used to bundle a large number of wires into a cable. The most common strap has a self-locking clip that keeps the cable in place once it is pulled tight (Figure 8-15). The tool in Figure 8-15(b) performs the same operation of tightening and cutting the strap as shown in Figure 8-15(a).

Clips are used when you wish to *anchor* a bundle of wires to a chassis or wall. To install a clip, first determine where to anchor the

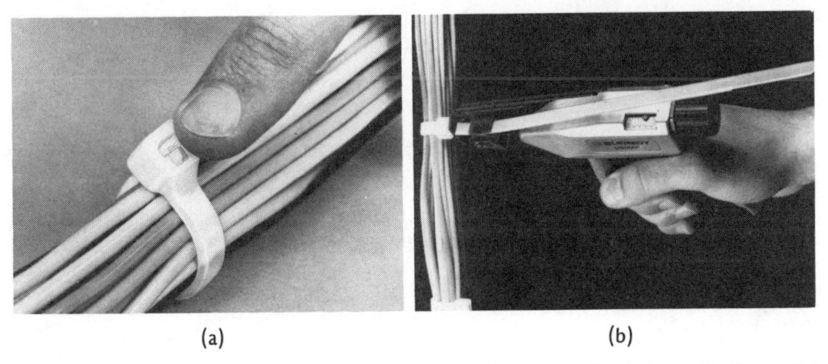

(a) (b)

Figure 8-15 Nylon, self-locking straps: **(a)** mechanical installation, **(b)** using a hand-operated tool for installation

bundle. At that point, attach the *stationary* part of the clip—a metal bracket—using a screw. Then clip the bundle of wires together with a spring latch which hooks onto the metal bracket (Figure 8-16).

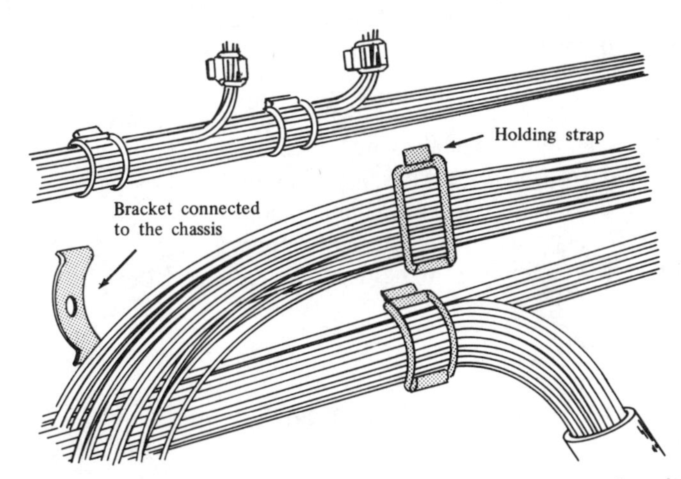

Figure 8-16 Using metal clips to connect cables to a chassis

KT's KEY TERMS TO REMEMBER

Anchor To secure firmly.

Impregnate To fill or saturate.

Jig Board A chassis arranged so that it can be used for temporary hookups for fast construction or dismantling.

Lacing Drawing together by means of ties or laces; tying together.

Plastic A highly pliable, synthetic material that is a poor conductor.

Prefabricated Already constructed in standard sections for ease of assembly.

Service Loop A small amount of extra wire left at a terminal connection so that the lead may be removed for service and resoldered to the terminal.

Splice A joining made by interweaving the strands.

Stationary Fixed, immovable.

Template A pattern used as a guide to make similar items.

TEST

1. Describe a wire harness.
2. How does a wire harness differ from a cable?
3. Name the two lacing techniques commonly employed in assemblies.
4. What is a breakout point and why do we have them?
5. Why do we use a template or harness board when making a wire harness?
6. List the steps required to put together a harness board.
7. What is the purpose of sleeving?
8. List and describe three types of tubing used as sleeving.
9. List four steps to follow in installing loose tubing.
10. What is the principal advantage of zipper tubing?
11. When do you apply a clip holder to a bundle of wires?

PROJECT 1: TYING LACE TO A DOWEL

Materials

⅜ to ½ inch wooden dowel and a quantity of lacing twine

Procedure

Correctly tie six spot ties on the dowel end; after these are checked by the instructor, tie a running lace with six ties on one end of the dowel, p. 98.

PROJECT 2: HARNESS MAKING

Materials

Harness board and the proper length of wire

Procedure

1. Use the harness board and wire supplies to assemble a harness that will meet the specifications of the instructor.
2. Be sure that your lacing ties are correct.

PROJECT 3: APPLYING SHRINK TUBING

Materials

1. Heat gun
2. Wire
3. Two in. shrink tubing

Procedure

Correctly place a 1 in. piece of shrink tubing around each end of the wire and shrink the tubing with the heat gun.

PROJECT 4: PLACING A BUNDLE OF WIRES TO FORM A CABLE

Materials

1. Several insulated conductors
2. 12 inch piece of plastic tubing
3. 24 inch lacing twine
4. Talcum powder

Procedure

Make a cable by placing the wires in the plastic tubing as described on page 106.

PROJECT 5: STRAPPING A CABLE WITH SELF-LOCKING STRAPS

Materials

1. Four nylon straps
2. Hand-operated strapping tool
3. Several insulated conductors

Procedure

1. Bundle the wires and tie them together with the hand-tightened nylon straps.
2. Add several more straps with the strapping tool.

9: Splicing Conductors

9.0 INTRODUCTION

Splicing is sometimes necessary to repair a break or lengthen a conductor. Also, sometimes a conductor is spliced in order to provide a service lead—a conductor connected to a device constructed of fine wires. Service leads provide durable connections and protect the fine wire from the pull and push which the connections must withstand. For example, in power transformers service leads are spliced to the fine transformer wire. Remember a good electrical connection must result from the splice.

9.1 PREPARATION FOR SPLICING

The following are some basic steps to follow to ensure that a good electrical connection is the final result of a splice.

1. Clean the surfaces of the conductors to be spliced.
2. After cleaning, tin the conductors.
3. After tinning, twist the conductors together to form a strong joint.
4. Heat conductors with a soldering iron and apply sufficient solder to form a solid electrical bond between them.

Some conductors have a varnish coating in addition to an insulating cover of cloth, plastic, or rubber. This must be removed with sandpaper or lacquer thinner in order to form a good electrical connection. Examples of splicing procedures are shown in Figure 9-1.

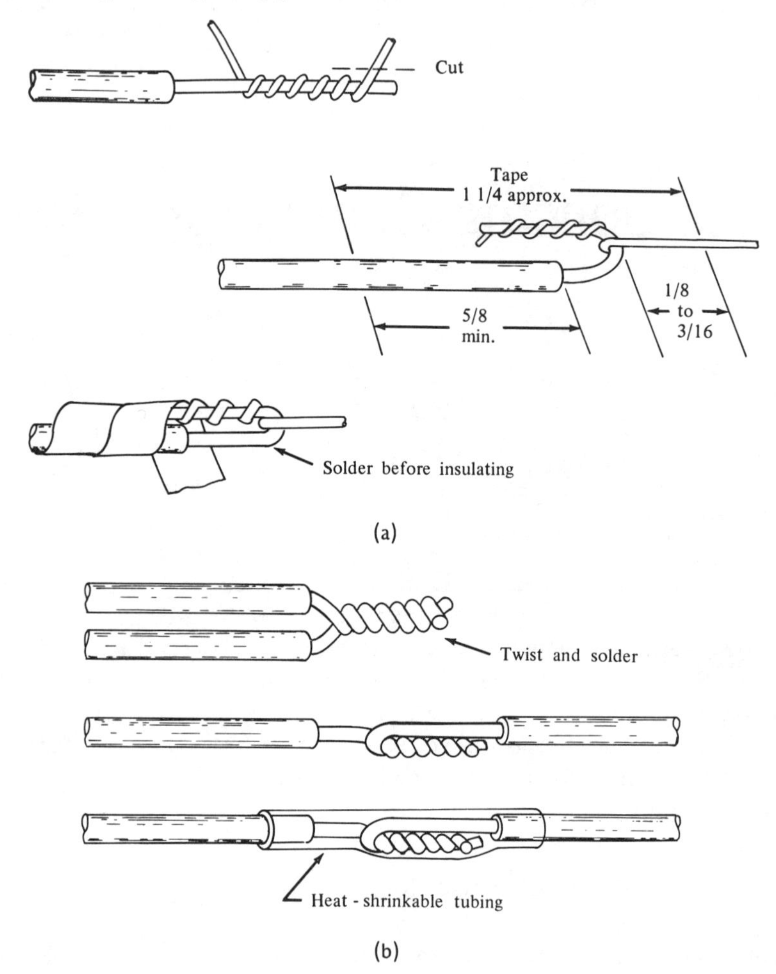

Figure 9-1 Splicing by twisting: **(a)** small to large wire, **(b)** same-sized wires

Heat-shrinkable tubing is an effective way to insulate the splice and is easy to use. The following is a review of the installation procedure:

1. Cut a piece of tubing the desired length.
2. Slip the tubing piece over one of the wires.
3. Complete the wire splice.

4. Slide the tube over the splice and heat. The tube will shrink and seal the bare splice (Figure 9-2).

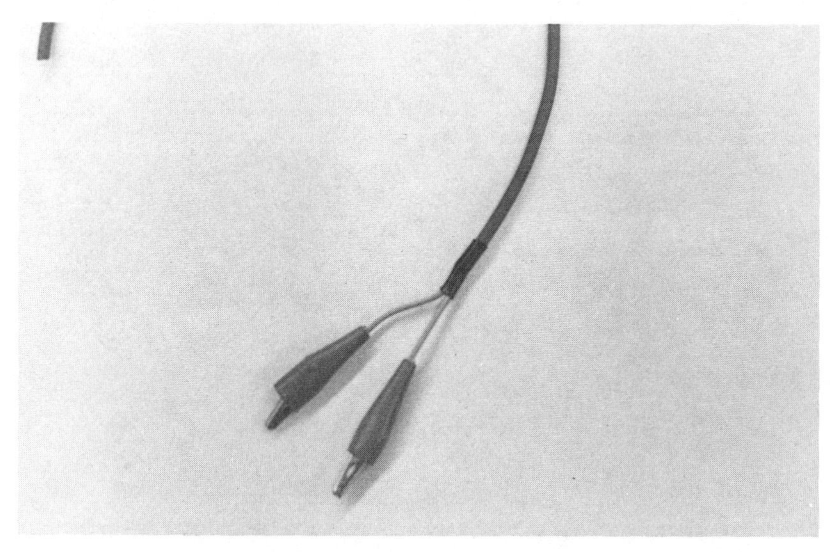

Figure 9-2 Example of heat-shrinkable tubing over a splice

From time to time, it may be necessary to mend broken wires by splicing. Large-diameter wires can be spliced by being twisted together; however, generally they are spliced by crimping (Figure 9-3). Small-diameter wires must be twisted together because crimping may nick them to the point where they become weak.

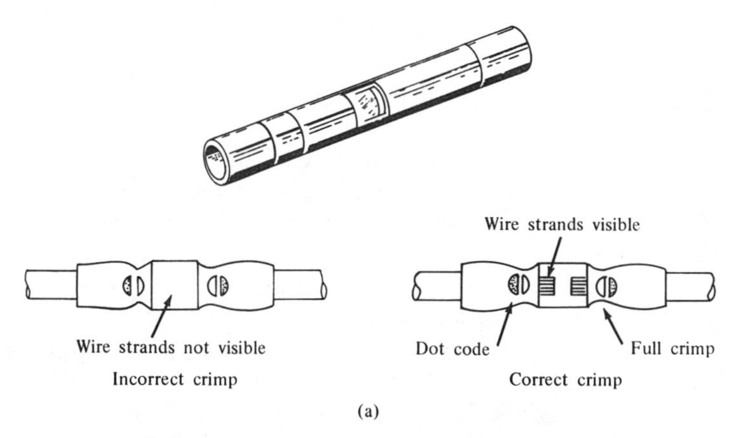

Figure 9-3 Splicing by crimping: **(a)** crimping tube

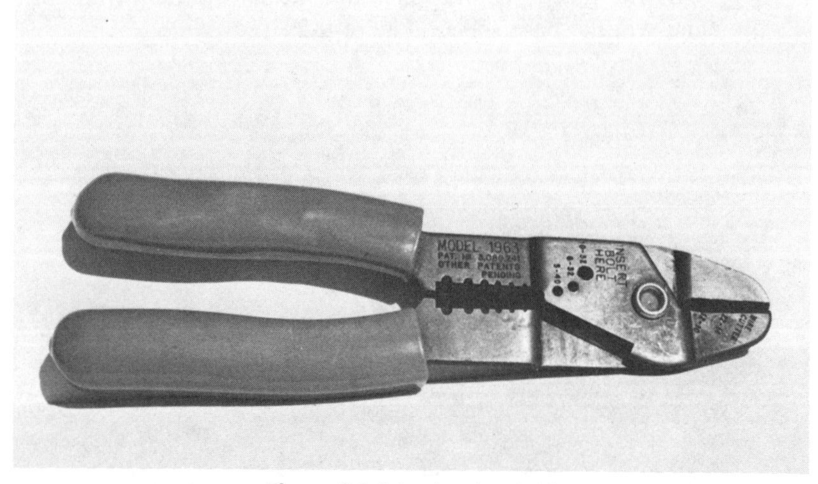

Figure 9-3 (b) crimping tool

To crimp wires together: (1) strip enough insulation off the wires so they fit into the crimping tube, (2) insert the wires into the tube, and (3) press the crimping tool around the tube to complete the splice. To determine whether the crimp is correct, check the center of the crimping tube to see whether wire strands are visible from each of the wires being spliced. If not, the crimp must be removed and the splice reworked.

TEST

1. Why would you splice a conductor?
2. List and describe the four basic steps required for a good splice.
3. Why must you use sandpaper or lacquer thinner to clean some wires?
4. What is the purpose of heat-shrinkable tubing?
5. When do we use crimping as a splicing technique?
6. If you had to splice two small wires, would you crimp them or solder them?
7. How do you know if a crimp has been performed correctly?
8. What is the effect of heating-shrinkable tubing?
9. If you had to splice two large-diameter wires, would you solder them or crimp them?
10. Why do you splice to a service lead?

PROJECT 1: SPLICING CONDUCTORS

Materials

1. Shrink cable tubing
2. 18-in. #20 insulated wire
3. 6-in. #22 wire
4. Heat gun

Procedure

Form two splices as shown in Figure 9-1. Have your instructor check the splices and then add the shrinkable tubing.

PROJECT 2: CRIMPING A SPLICE

Materials

1. Crimping tool
2. 12-in. #20 insulated wire
3. Wire strippers
4. Two crimping tubes for #20 insulated wire

Procedure

1. Strip the wire the proper length to fit into the crimping tube.
2. With the crimping tool, crimp the tubes over the wires as shown in Figure 9-3.
3. Have your instructor check the splice.
4. Add shrinkable tubing over the splice if the crimping tubes are not the insulated type.

10: Wire Wrapping

10.0 INTRODUCTION

Wire wrapping (also called solderless wrapping) is the *technique* of connecting wires to terminals without soldering. Wire is wrapped tightly around a sturdy metal *stud* resulting in a connection that is mechanically and electrically acceptable. Remember to learn the KT's at the end of the chapter.

10.1 WIRE-WRAP MATERIALS AND TOOLS

Two materials are used: wire-wrap terminals and wire. Wire-wrap terminals (sometimes called posts or pins) are usually made of a copper alloy and are square or rectangular, ranging in length from approximately 0.3 to 0.6 in. The side dimensions are generally 0.025 to 0.045 in. Figure 10-1 shows how a wire-wrap terminal is soldered to an insulation board anchor.

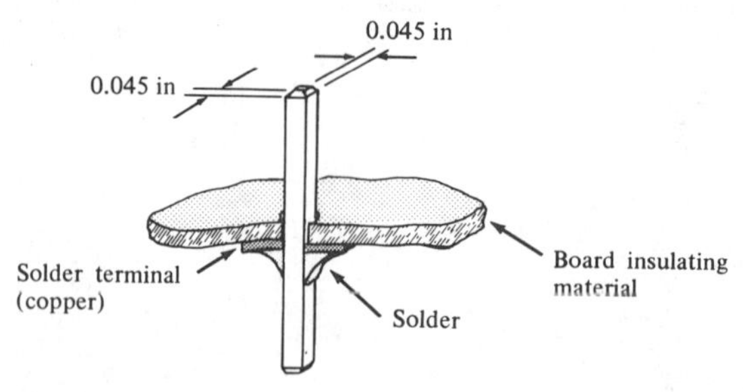

Figure 10-1 A wire-wrap terminal soldered to an insulation board anchor

The wire used to produce good wire-wrapped connections ranges in size from #20 to #26. The electrical demands of the circuit dictate the size of the terminal and wire to be used. To ensure good connections, both terminals and wire are plated with either a silver or gold alloy.

10.2 WIRE-WRAP PROCESS

The connections or wrapping can be done manually or with automatic tools. Some hints to remember are:

1. Remove insulation from the wire without *nicking* the wire.
2. Position the wire so the normal pull from its connection point does not unwind the wire.
3. Use only nonmetallic, blunt instruments for dressing wires. An automatic wire-wrap tool is shown in Figure 10-2.
4. To remove a winding, use a dewrapping tool as shown in Figure 10-3.
5. For proper wire-wrapping results see Figure 10-4. *Note:* Figure

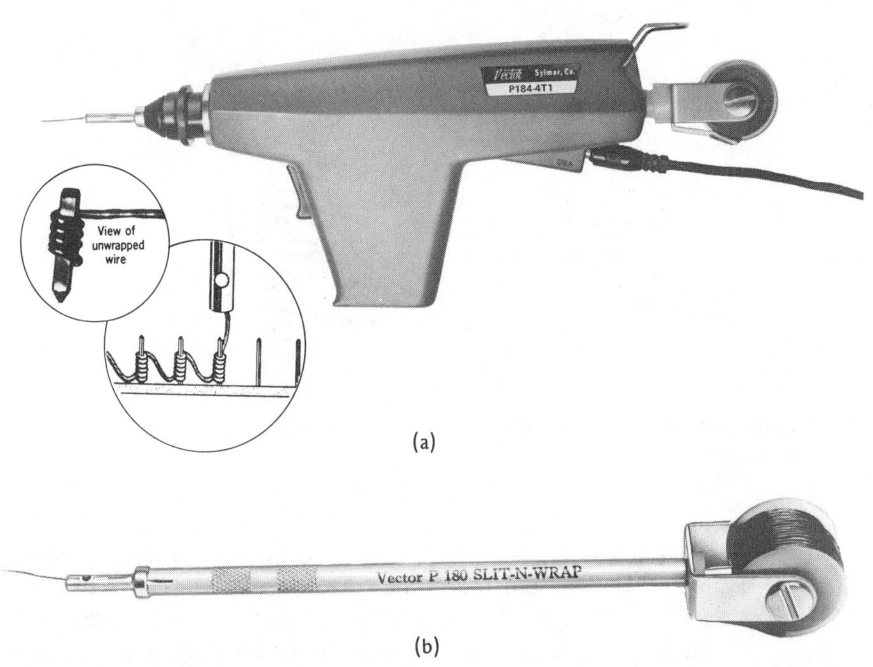

(b)

Figure 10-2 A wire-wrap tool: **(a)** a battery-operated gun, **(b)** the gun in operation

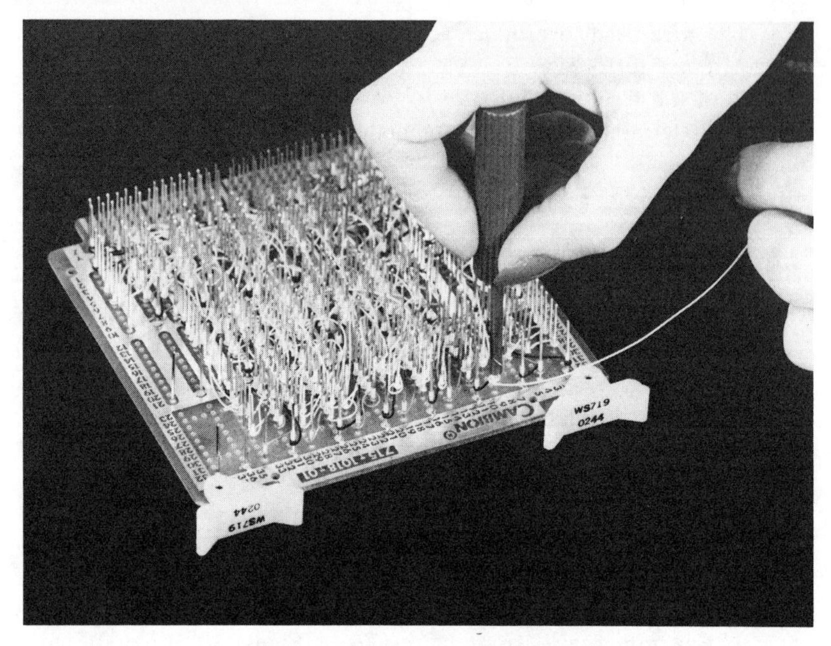

Figure 10-3 A dewrapping tool

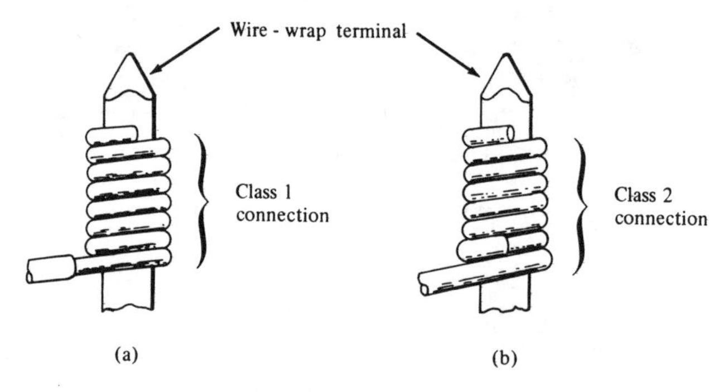

Figure 10-4 Proper wire wrap: **(a)** class 1 connection, **(b)** class 2 connection

10-4(a) shows a class 1 (or type a) connection where the wire insulation does **not** touch the post. Figure 10-4(b) shows a class 2 (or type b) connection where the wire insulation touches the post for one full turn. Standard wrap is 5-8 turns.

Following are some don'ts to remember (Figure 10-5).

1. Don't leave fewer than five turns.
2. Don't overlap wraps.
3. Don't leave pigtails.
4. Don't overwrap.
5. Don't spiral or open wrap.

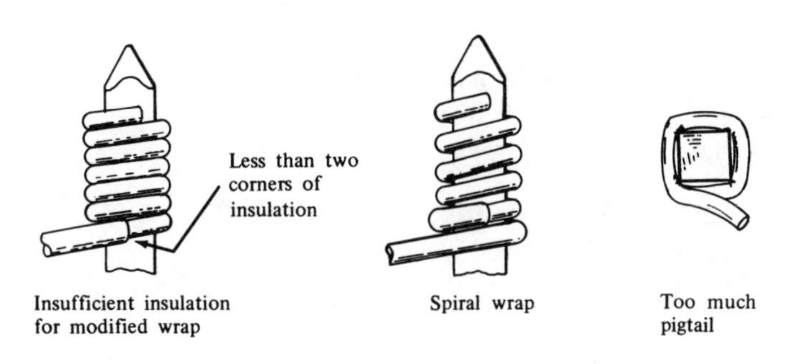

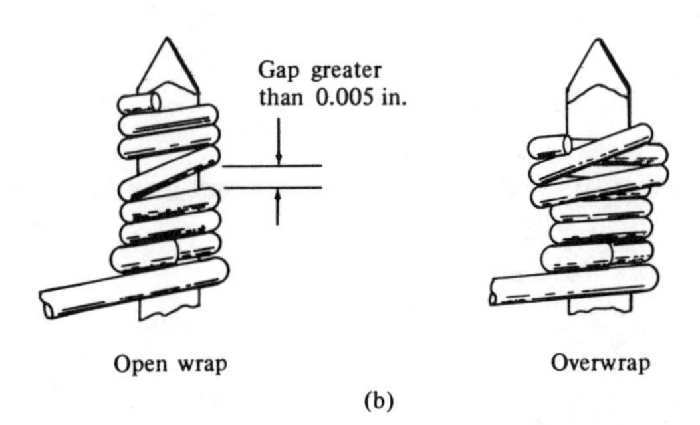

Figure 10-5 Samples of wire wrapping: **(a)** acceptable wrap, **(b)** unacceptable wrap *(Courtesy of Gardner Denver Company)*

KT's KEY TERMS TO REMEMBER

Nicking Making a small notch, groove, or dent; in wire, this may weaken it.
Stud An upright post used as an anchor for wires.
Technique The systematic procedure by which a task is performed.

TEST

1. Describe wire wrapping.
2. Why is wire wrapping used?
3. What are the materials employed in wire wrapping?
4. What materials are used to coat the wire and terminals of a wire-wrap connection?
5. Why do we use nonmetallic, blunt instruments to dress the wires?
6. When wrapping, is it a good idea to overlap wrappings?
7. Is a pigtail left for added strength?
8. If we use an automatic wire-wrap tool, is it acceptable to use nicked wire?
9. What is the name of the hand tool used to remove a wire-wrap connection?

PROJECT: WIRE WRAPPING

Materials

1. Circuit board with metal wire-wrap studs installed.
2. Wire-wrap tool and wire.
3. Dewrapping tool.

Procedure

1. Make 10 proper wire-wrap connections and have the instructor check your work.
2. Unwrap the wire and clean the board.

11: Flat-Wire Cable

11.0 INTRODUCTION

Flat-wire or ribbon cable (Figure 11-1) is composed of wires bound together to form a flat conductor. This type of cable is inexpensive to manufacture, lightweight, flexible, and easy to use in either small or large equipment. Figure 11-2 shows some of the applications of flat-wire cable. Remember to learn the KT's at the end of the chapter.

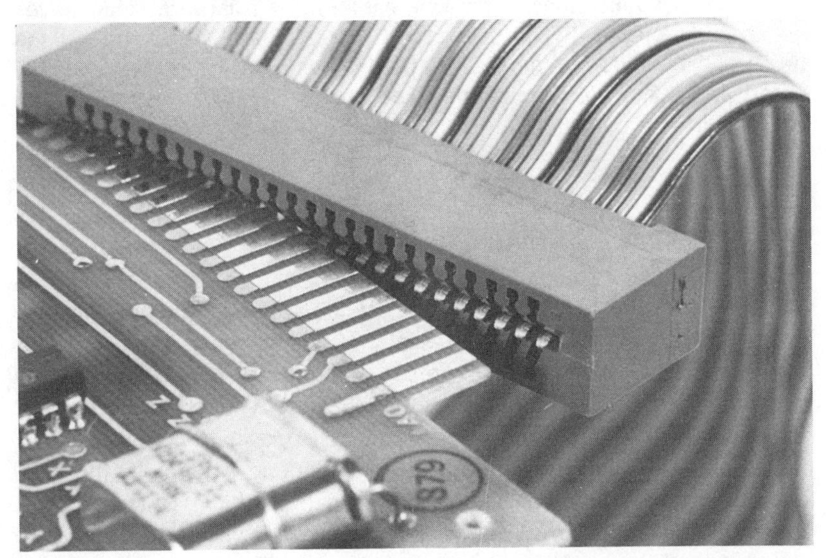

Figure 11-1 A flat-wire cable *(Courtesy of Watkins–Johnson, Inc.)*

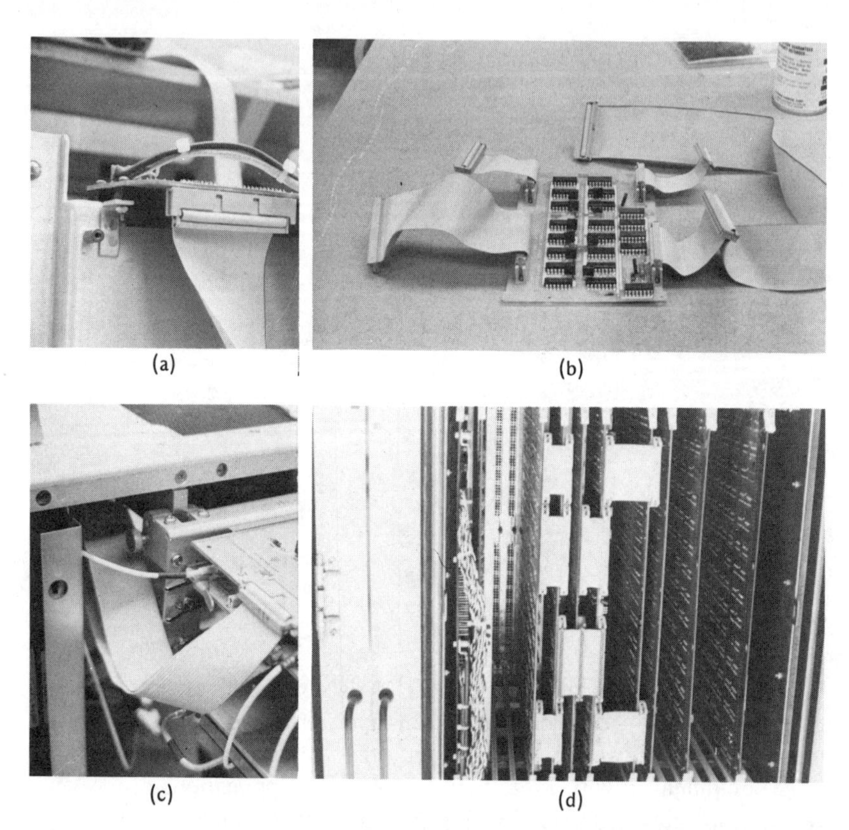

Figure 11-2 Applications of flat-wire cable: **(a)** the flat-wire cable serves as an input-output (I/O) connector between the light-emitting diode (LED) and the information source, **(b)** PC board with flexible cable between board and communications-type connector, **(c)** as a socket connector, **(d)** as a jumper between boards *(Courtesy of Watkins–Johnson, Inc.)*

11.1 STRIPPING FLAT-WIRE CABLE

Stripping flat-wire cable requires special tools and techniques that are not necessary for stripping individual wires. The most common stripping technique, called "cold stripping," can be done in several ways using different kinds of tools. Figure 11-3 shows cold stripping *parallel* to the conductors. The fiber wheels rotate in opposite directions, removing insulation from both sides. The *buffing,* done parallel to the conductors, creates *friction* and melts the insulation which is then brushed away by the buffer brushes. When the buffing wheels are correctly separated, all the insulation is removed leaving the conductors clean.

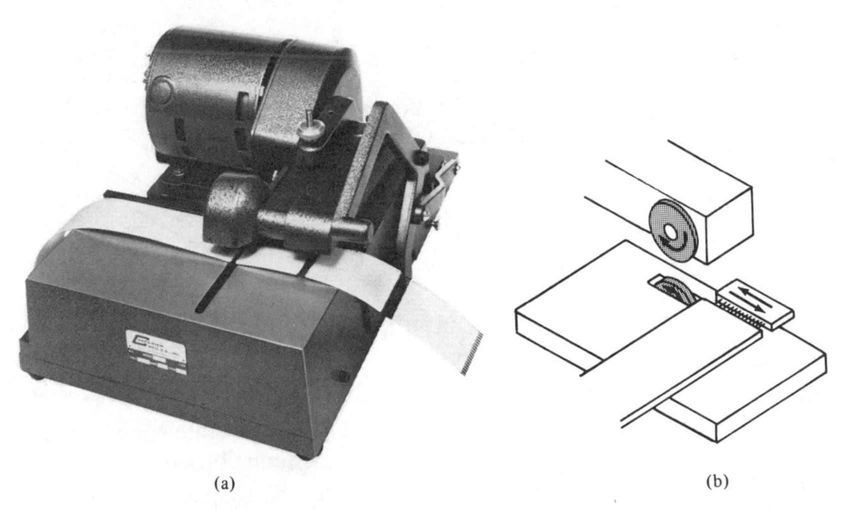

Figure 11-3 Cold stripping flat-wire cable with a friction-wheel stripper: **(a)** machine, **(b)** details of operation. *(Courtesy of Carpenter Manufacturing Company)*

Details of the cold-stripping method are shown in Figure 11-3b. The two buffer wheels turn in the same direction to remove insulation. The insulation is heated by friction and brushed away. The advantage of this kind of stripper is that it can be used to clean a portion of the cable in the center.

Figure 11-4 shows the proper use of a *hand plane* for stripping cable. The hand plane is a device similar to the block plane used for wood working. Take care that the blade does not *protrude* more than is needed to remove the insulation.

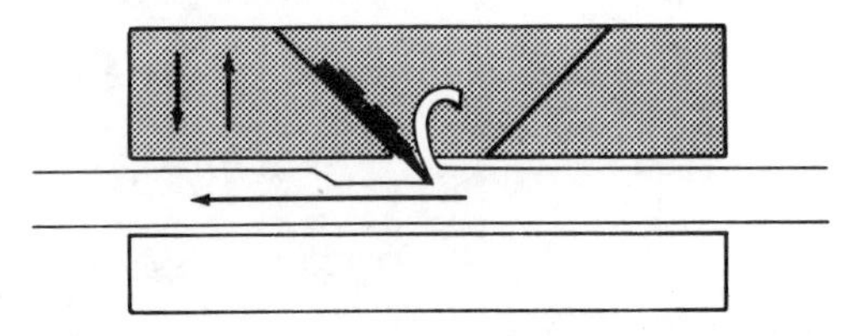

Figure 11-4 Cold stripping with a hand plane

Cold stripping with a blunt knife edge is shown in Figure 11-5. The knife is electrically heated and, when it is hot enough to melt the insulation, the cable is pulled past the knife. Then the insulation can be removed by buffing or an appropriate solvent.

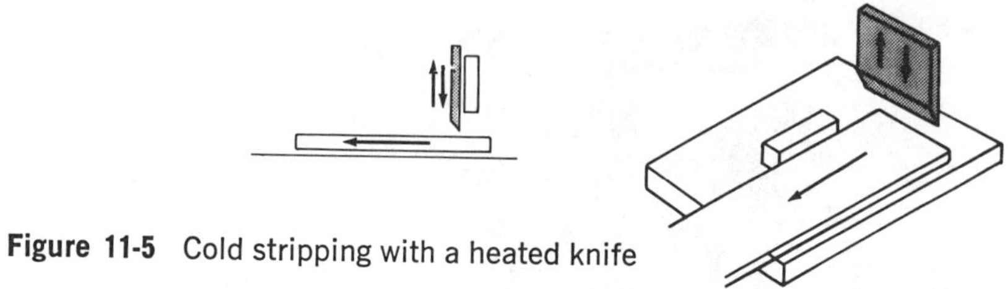

Figure 11-5 Cold stripping with a heated knife

11.2 FLAT-WIRE CABLE TERMINATIONS

The examples in Figure 11-6 show the most commonly used methods for making flat-wire cable terminations. These are female connectors clamped or soldered to one-sided printed-circuit boards. Whatever termination method is used, the cable must first be stripped, tinned, and anchored to the connecting unit, if the cable wires are not pre-tinned.

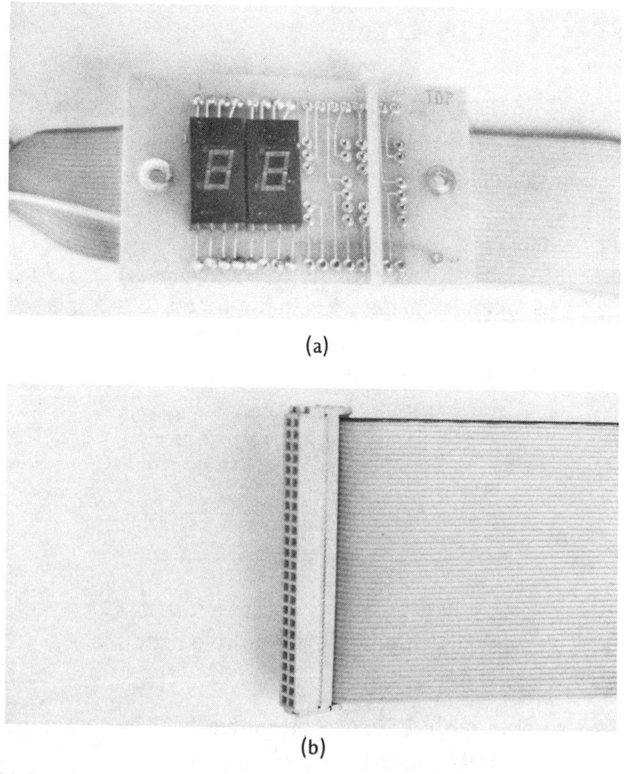

(a)

(b)

Figure 11-6 Samples of flat-wire cable terminations: **(a)** soldered method, **(b)** clamped method

KT's KEY TERMS TO REMEMBER

Buffing To rub or polish to a smooth finish.
Friction The resistance between two moving bodies on contact; friction causes heat.
Hand Plane A handheld tool containing a knife that is used to shave or smooth a surface.
Parallel Alongside and everywhere equidistant.
Protrude To extend beyond a certain point.

TEST

1. What is a flat-wire cable?
2. How is insulation removed from the middle of a flat-wire cable?
3. Discuss two methods for removing insulation from the end of a flat-wire cable.
4. How is ribbon cable terminated?

PROJECT: FLAT-WIRE CABLE

Materials

1. Length of flat-wire cable
2. Cable-stripping tool
3. Flat-wire cable termination connector

Procedure

1. Properly strip the flat wire and have the job approved by the instructor.
2. Connect the cable to the termination connector.

12: Joining Techniques

12.0 INTRODUCTION

Joining metals by soldering was discussed in detail in Chapter 7. In this chapter we will discuss other methods of joining metals to metals, metals to nonmetals, and nonmetals to nonmetals. Remember to learn the KT's at the end of the chapter.

12.1 RESISTANCE WELDING

Welding is a process whereby heat and pressure are applied to two metals to be joined. In the resistance-welding method the two metals are squeezed together between two *electrodes* by a welding machine such as the spot welder shown in Figure 12-1 (a).

Details of the electrode contact are shown in Figure 12-1(b). A good weld requires the heat to be concentrated at the point of the junction of the metals. For this reason, metals with high resistance (low conductivity), such as nickel and nickel alloys, are welded with electrodes that have a low resistance. Metals with a low resistance (high conductivity), such as copper, copper alloys, gold, and gold alloys, are welded with electrodes that have a high resistance.

Figure 12-2 shows details of some acceptable and unacceptable welds of component leads to flat surfaces. Many component leads are made of alloys with a special *noncorrosive* coating that will sometimes cause problems with the weld. This means that leads of the same diameter may require different machine settings for good results. The

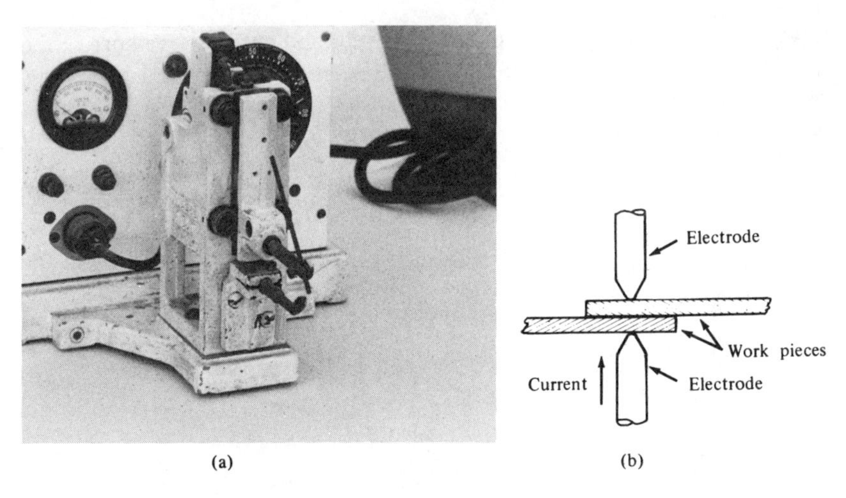

Figure 12-1 **(a)** spot welder, **(b)** details of electrodes

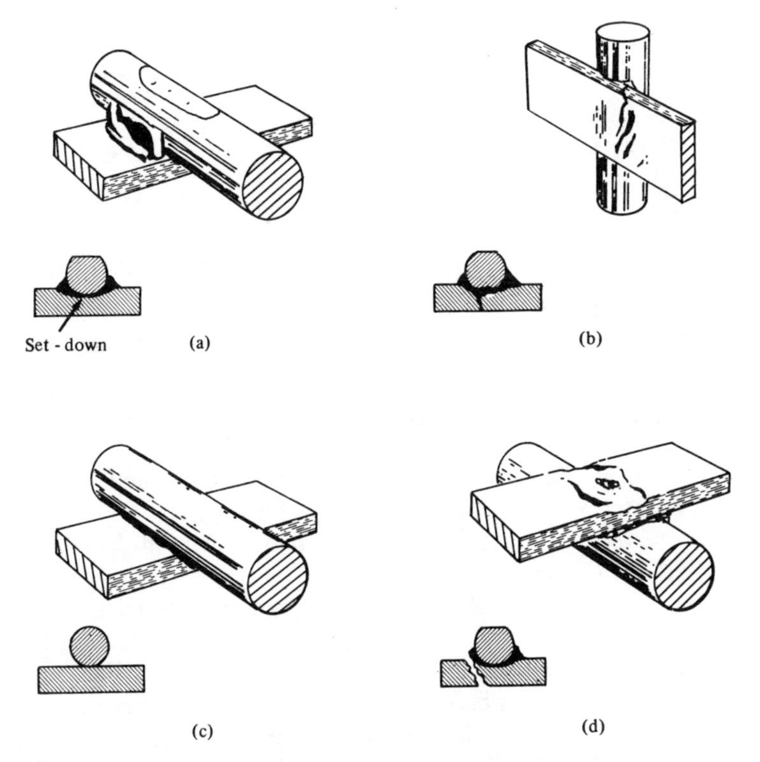

Figure 12-2 Acceptable and unacceptable welds: **(a)** good weld, **(b)** cracked weld, **(c)** insufficient weld, **(d)** burned hole

machine settings are usually pressure, dwell, and current. Pressure is the amount of force between the tips of the electrodes measured in *pounds per square inch* or *grams per square centimeter.* Dwell is the amount of time current is allowed to flow through the electrodes. Finally, current is the level of energy that flows through the electrodes to heat the metals. Samples of proper component welds are shown in Figure 12-3.

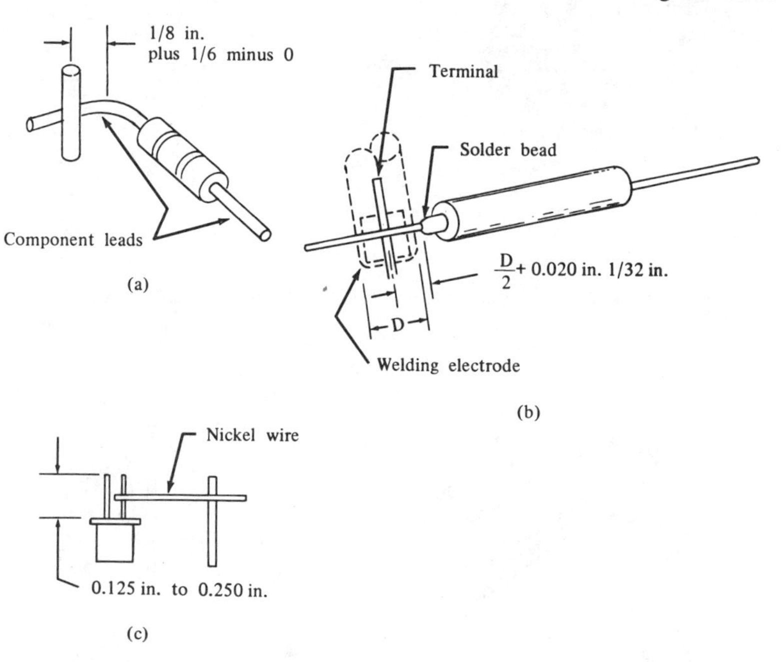

Figure 12-3 Proper welds of lead extension ribbons to component leads: **(a)** maximum bend, **(b)** terminal connection, **(c)** transistor connection

Some general rules for welding components are:

1. Do not weld one component to another component.
2. Do not weld a component to a stranded wire.
3. As much as possible, make welds that create a 90° angle to the *ribbon* (Figure 12-4).
4. When welding more than one ribbon to a terminal or lead, separate each ribbon by at least the width of the ribbon (Figure 12-5).
5. Form the weld ribbon around the component lead so that the ribbon is inside the bend (Figure 12-6).
6. Don't bend the ribbon too close to the component lead (Figure 12-7).

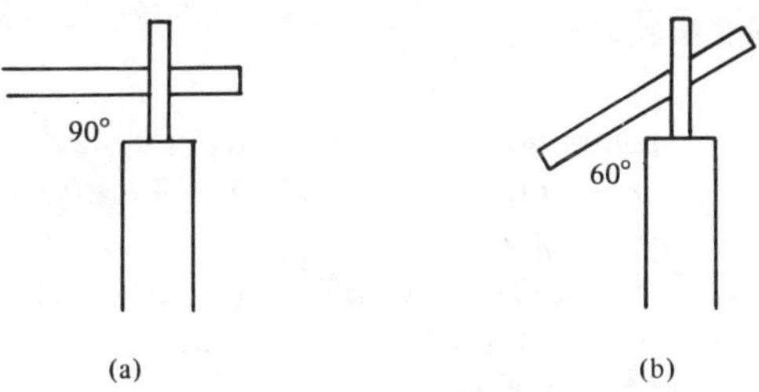

Figure 12-4 Attaching the ribbon to the component lead: **(a)** good, **(b)** poor

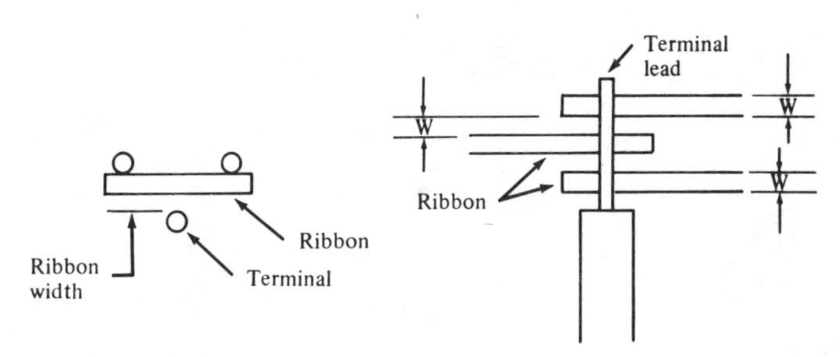

Figure 12-5 The distance from the weld ribbon to a terminal or other ribbon should be equal to at least the width of the ribbon

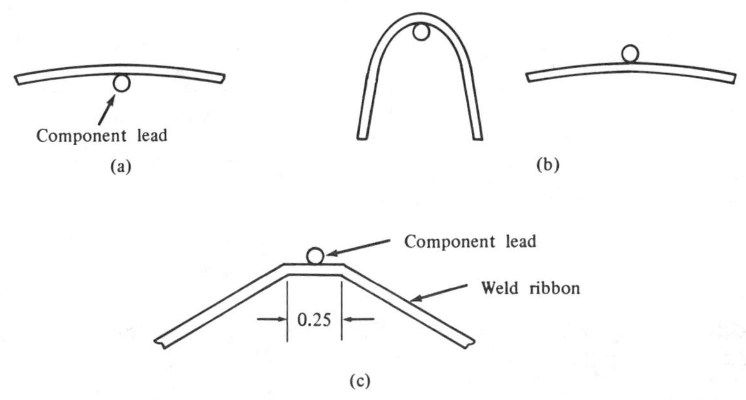

Figure 12-6 Forming the ribbon around the terminal or lead: **(a)** acceptable, **(b)** unacceptable, **(c)** alternate method

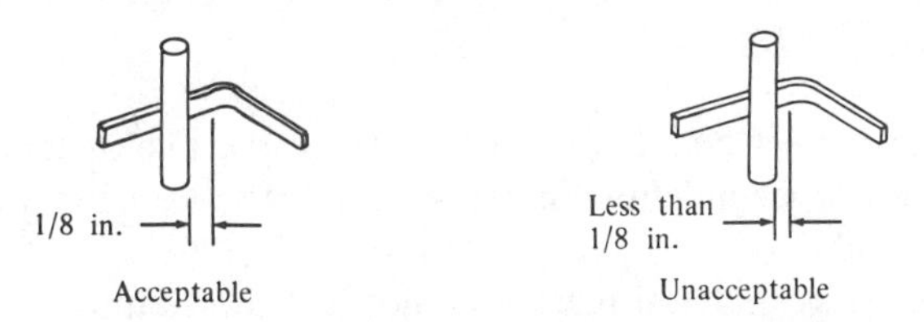

Figure 12-7 The ribbon should not be bent too close to the lead

7. When two components are connected by a rectangular ribbon it is desirable to put a twist in the ribbon between the components (Figure 12-8).
8. There should be very little offset or overlap when two ribbons are welded in line (Figure 12-9).
9. Visually inspect each weld.

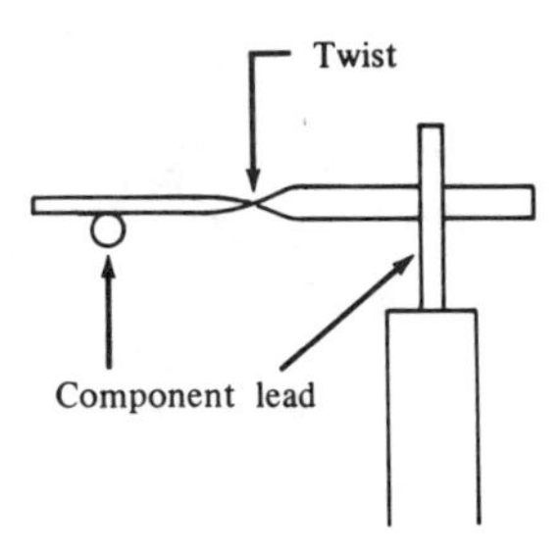

Figure 12-8 Placing a twist in the ribbon when two components are being connected

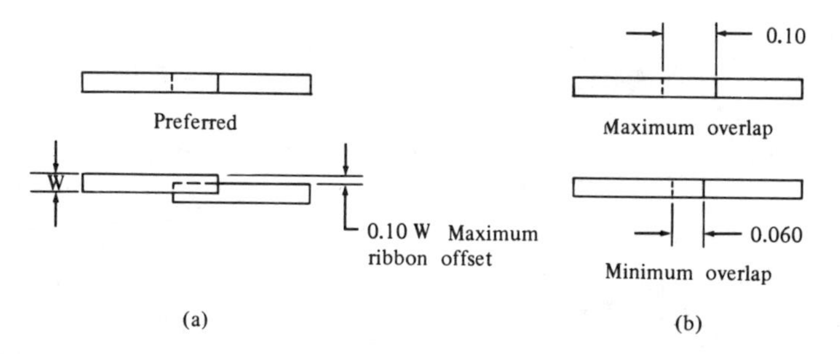

Figure 12-9 Offset and overlap: **(a)** acceptable and unacceptable offset, **(b)** maximum and minimum overlap

Each weld should be inspected, compared to Figure 12-10, and rejected for any of the defects shown in Figure 12-11. Notice the following in Figure 12-10:

1. The correct quantity of melt between the two materials.
2. The proper indentation (set-down) of material at the point of the weld.
3. The base material is welded, not just the coating.

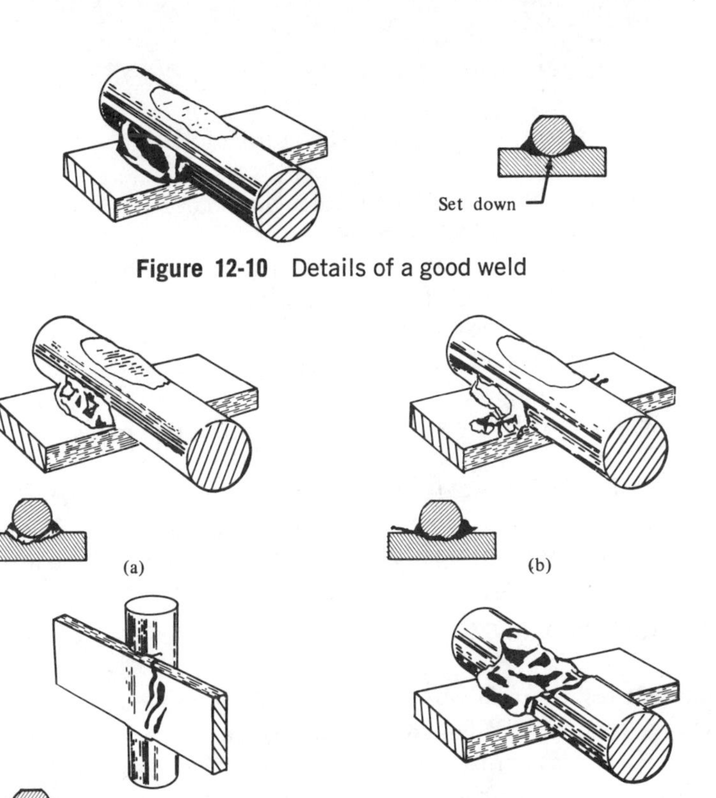

Figure 12-10 Details of a good weld

Figure 12-11 Unacceptable welds: **(a)** blow hole, **(b)** weld splatter, **(c)** cracked weld, **(d)** burned weld, **(e)** burned hole, **(f)** insufficient weld, **(g)** miscentered leads on electrodes

Defects shown in Figure 12-11 are:

(a) Blow hole—any hole greater than 0.003 in. in the fused area between the welded parts.
(b) Weld splatter—whiskers of metal protruding from the weld. Removing the whiskers will not make the weld acceptable.
(c) Cracked weld—a crack in or near the weld.
(d) Burned weld—when a welded piece is overheated so that its size is reduced by 30 percent.
(e) Burned hole—a hole of any size through either metal at the weld.
(f) Insufficient weld—no flow of melt around weld joint.
(g) Indication of miscentered leads on electrodes—visible indentation from the edge of the electrode on either of the metals.

12.2 HOT-MELT RESINS

Hot-melt resins are *thermoplastic* materials used for *adhesive* bonding, potting, coating, and holding things together. Hot-melt glues may be applied by spray, brush, dip, or heat gun. The most widely used dispensers are self-contained, slug-loading guns, such as illustrated in Figure 12-12. The glue slugs are placed in the gun and applied to the area to be filled or glued through heat and pressure. On the production line the gun may be fed by a hose from a pot.

The advantages of adhesives for product assembly are strength, economy, simplicity, and ability to bond different materials to each other.

Figure 12-12 A hot-melt adhesive gun

Although adhesives are excellent bonding materials, there are some drawbacks to their use.

1. Adhesives are sensitive to the surface conditions of the materials to be bonded. Special preparation of the surfaces and special handling may be required.
2. Some adhesives do not hold well when exposed to low temperatures, high humidity, severe heat, chemicals, water, and so forth.
3. Some adhesive solvents present a hazard to workers. Special ventilation requirements may be required.

12.3 JOINT DESIGN

Joints are designed to be used with structural adhesives. Selecting the appropriate joint is largely a matter of common sense and experience, but the goal should be to achieve as even a load distribution as possible so that the bonded area shares equally. Several types of joints are shown in Figure 12-13. The simple lap joint shown in Figure 12-13(a) is offset which may result in the joint peeling or twisting under a load. The tapered, single lap joint is better than the simple lap joint in that the tapered edge allows some bending of the joint under load. The joggle lap joint is better than either the simple lap joint or the tapered, single lap joint and can be formed easily with basic metal-working tools. The double-butt lap joint gives a more uniform load distribution than either the simple, tapered, or joggle joints but it requires machining that is difficult with thin metals.

Angle Joints

Figure 12-13(b) shows methods of making angle joints. Angle joints are subject to peel and cleavage stress depending on the gauge of the metal.

Butt Joints

Typical butt joints are shown in Figure 12-13(c). A straight butt joint has poor resistance to cleavage. The four butt joints shown have almost equal strength and should be selected by the machining process available. These types of joints form a pocket for the adhesive between the materials being bonded.

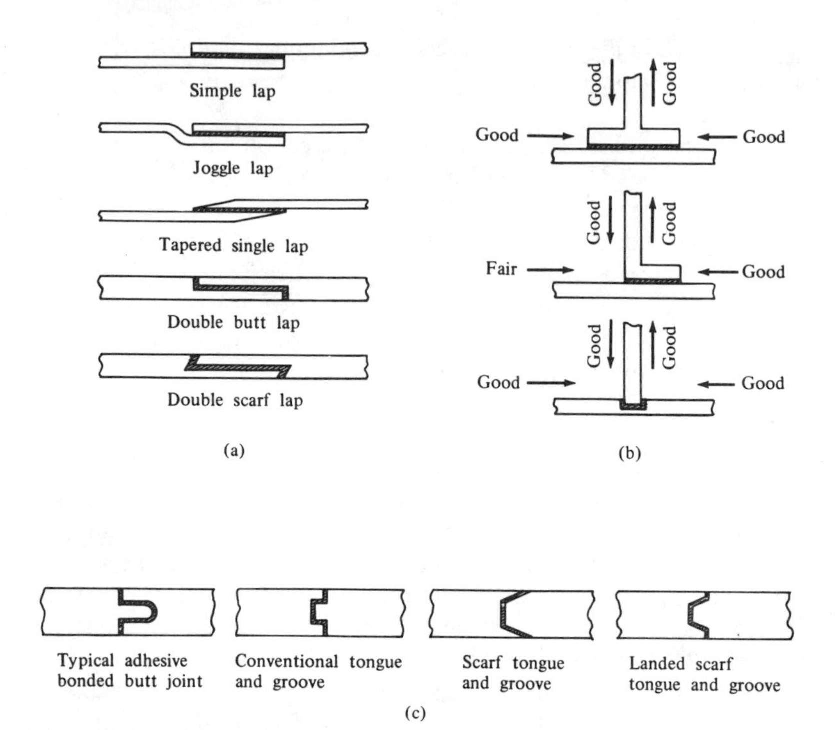

Figure 12-13 Samples of joints: **(a)** lap joints, **(b)** angle joints, **(c)** butt joints

Other Types of Joints

The T joint and the overlap slip joint (Figure 12-14 a) are typical of the joints used in bonding cylindrical parts such as tubing, bushings, and shafts. Adhesives provide a strong joint that is neater looking than a weld and is free of the distortion caused by high welding temperatures.

There are a number of corner joint designs. Corner joints used on sheet metal include the right-angle butt joint, slip joint, and right-angle support joint (Figure 12-14 b). Corner joints used where rigidity is necessary include the end-lap joint, mortise and tenon joint, and mitered joint with spline (Figure 12-14 c). Adhesives are used to fill the voids between the materials.

Stiffener joints are used to prevent deflection and flutter of thin metal sheets (Figure 12-14 d). Stiffening sections such as T-sections, hat sections, and corrugated backing can be used depending on the desired stiffness.

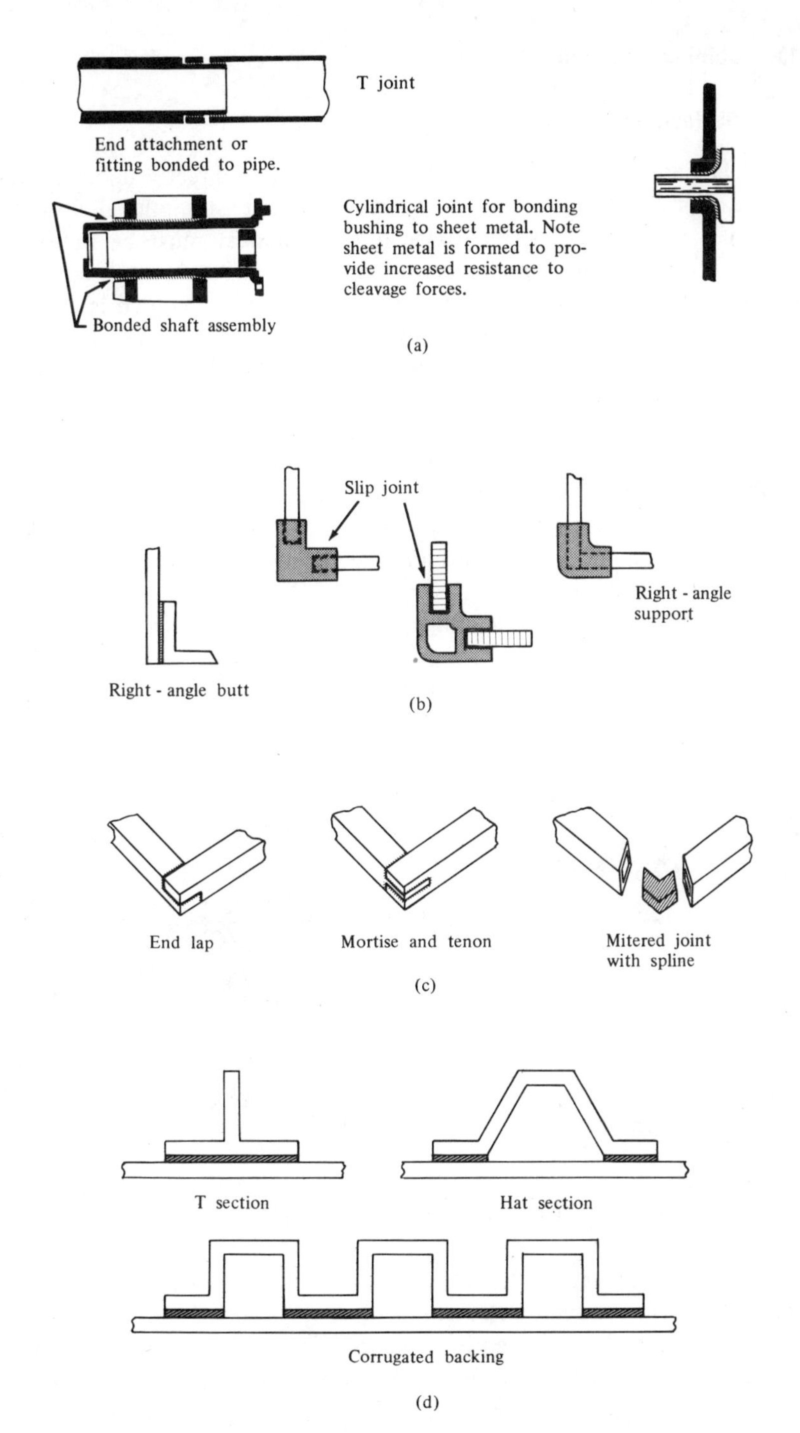

Figure 12-14 Other joints: **(a)** T joint on cylindrical parts, **(b)** corner joints—sheet metal, **(c)** corner joints—rigid, **(d)** stiffener joints

Surface Preparation

The amount of surface preparation depends directly upon the required bond strength. For maximum bonding strength all surface materials such as paint, oil, oxide films, and dust must be removed. There are three basic methods of removing contaminants: chemical cleaning, abrasion, and degreasing. Chemical cleaning, which is popular for preparing metals, includes treatments that etch the metal surface so that it will make a very strong bond. Whenever possible, chemical cleaning is the preferred method.

Abrasion cleaning includes sandblasting with fine sand on surfaces that are thick enough to resist distortion. Other abrasion cleaning methods are vapor honing and using finishing materials.

Degreasing may be used when maximum adhesive strength and outdoor use are not critical. Surfaces are cleaned with either a hot alkali solution or volatile solvent.

Regardless of the cleaning method used, the surface must be coated immediately with a proper coating to prevent oxide formations. Figure 12-15 shows some of the many uses of adhesive bonding.

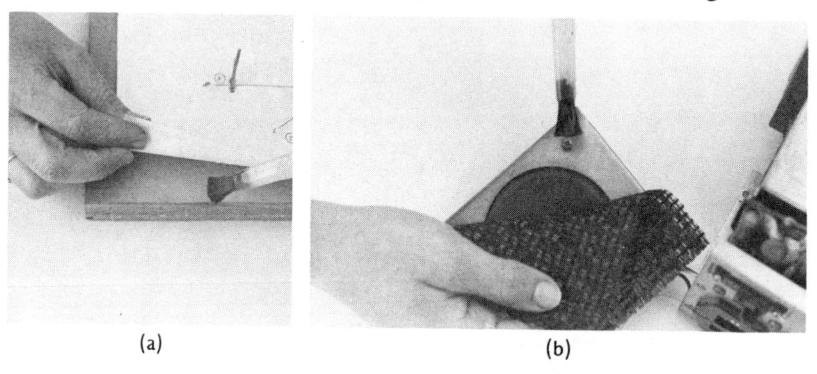

(a) (b)

Figure 12-15 Use of adhesives: **(a)** gluing speaker to grill cloth, **(b)** gluing wiring cable drawing to board.

12.4 MICROINTERCONNECT BONDING SYSTEMS

Transistors and integrated circuits (IC's) must have wire leads bonded from their semiconductor elements to a metal frame for connection to external circuits. The conventional method is to weld an individual wire to each element, one wire at a time. For example, the 16 wires in Figure 12-16 require 32 individual bonding operations. The microinterconnect bonding system eliminates the individual operations. Figure 12-17 is a graphic representation of inter-lead bonding where an IC is

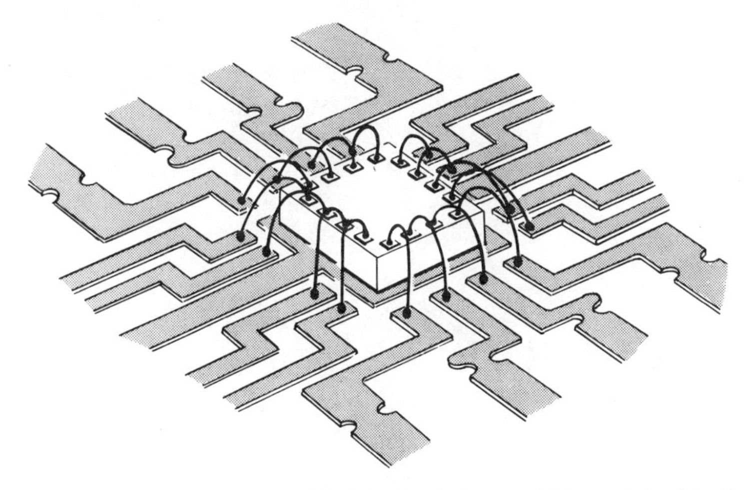

Figure 12-16 Individually welded leads between IC's and inside frame leads

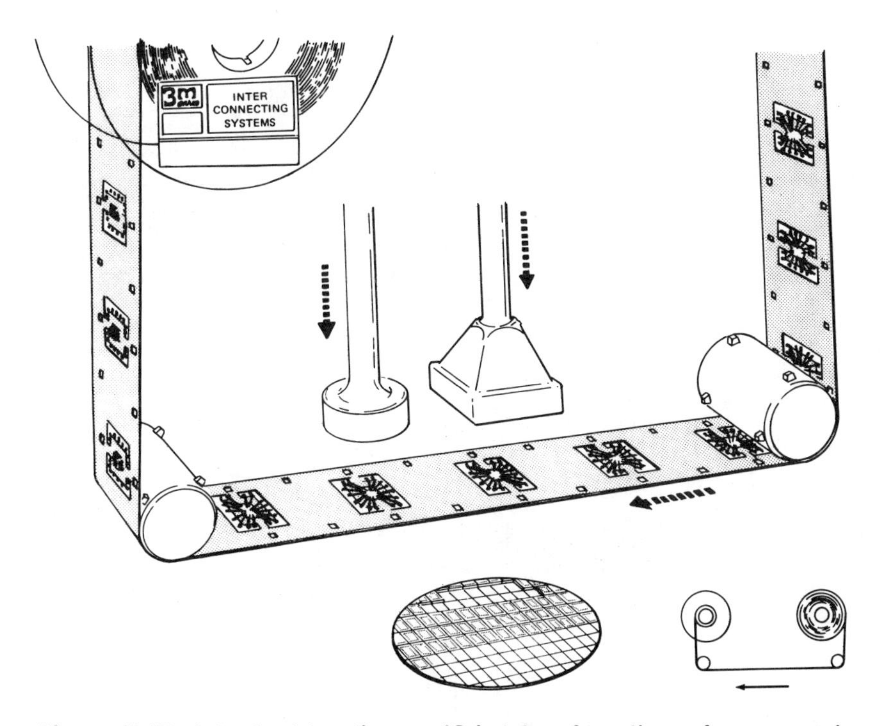

Figure 12-17 Inter-lead bonding: an IC is taken from the wafer array and bonded to each microinterconnect on a strip of beam tape

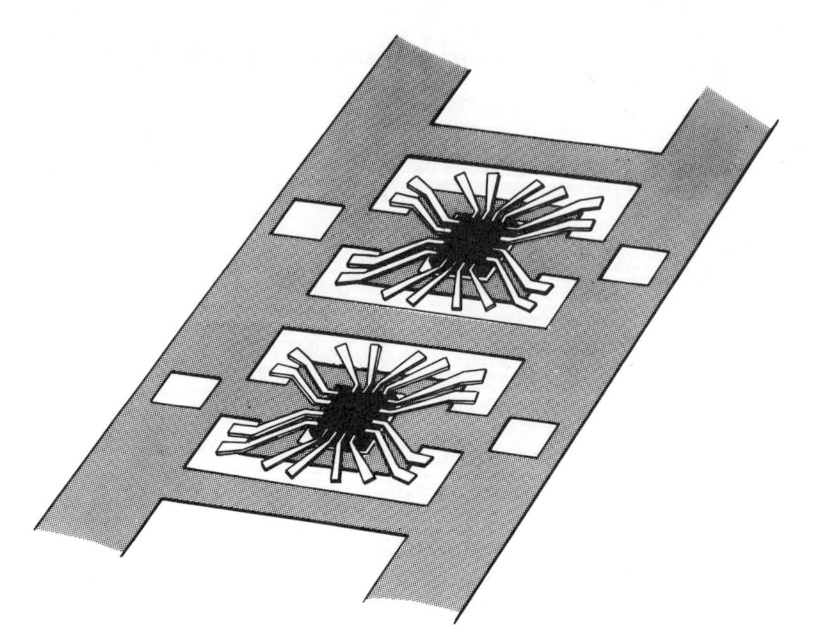

Figure 12-18 IC after being bonded to a microinterconnect strip

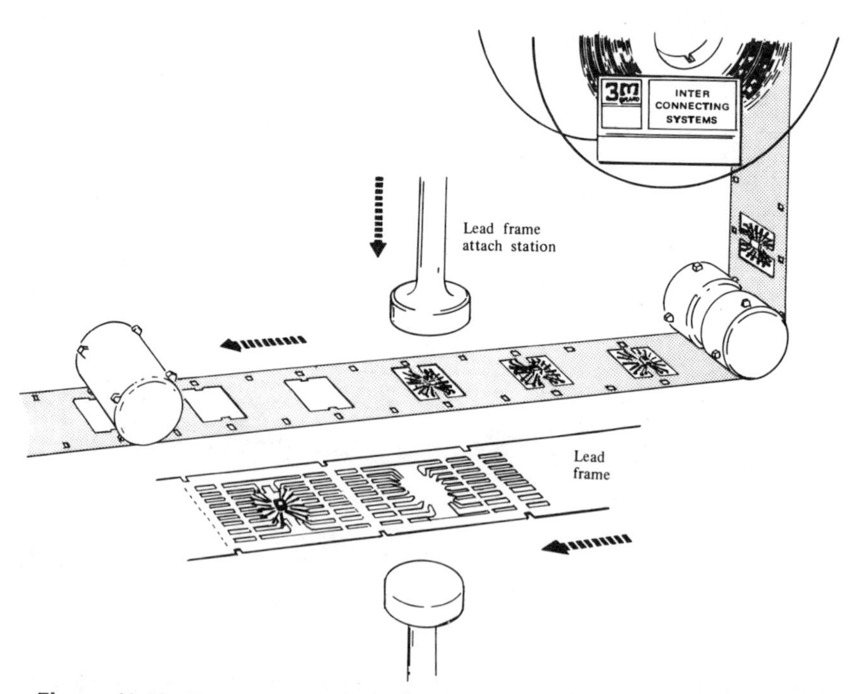

Figure 12-19 Bonding microinterconnects to the outer frame of IC's

taken from a wafer array and bonded to a microinterconnect on a strip of beam tape. The bond is made at the die attach station and then optically checked at the optical test station. Figure 12-18 shows how the bonding looks after this step. The holes at the tape edges are used to align the outer leads when they are bonded to the IC (Figure 12-19). The final bonding to the outer frame appears as shown in Figure 12-20(a)—24 inter and outer leads were bonded in two easy steps. The finished IC package is shown in Figure 12-20(b).

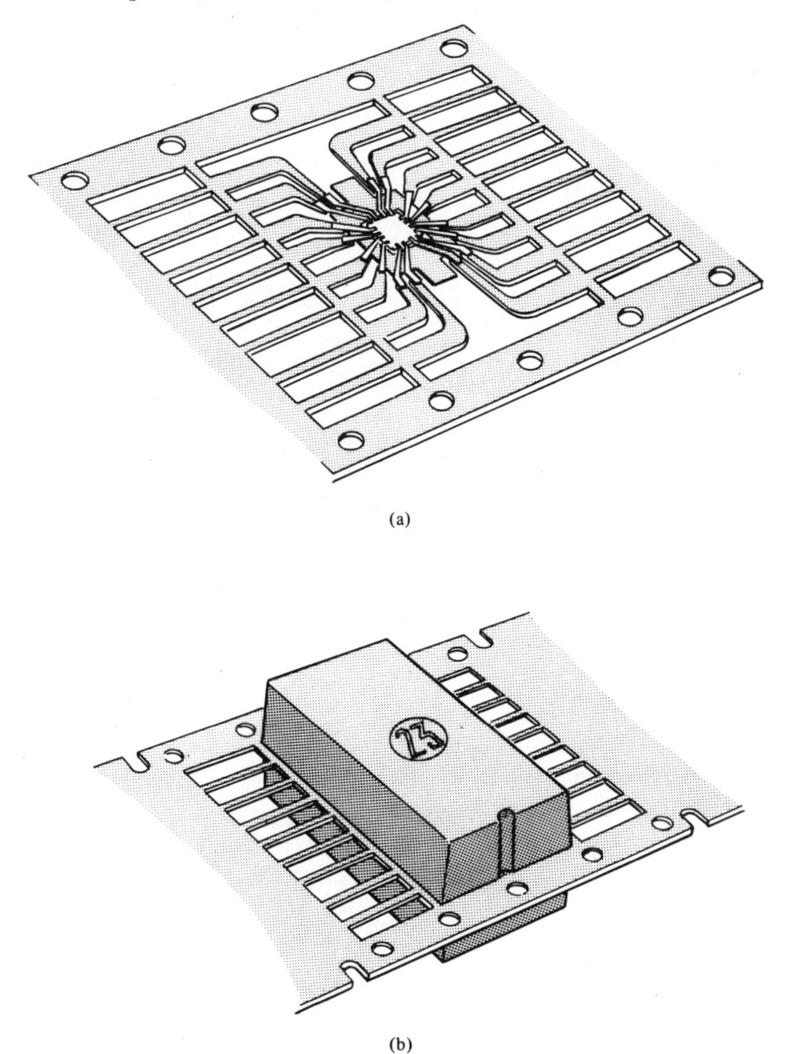

Figure 12-20 **(a)** IC with outer frame bonded to the microinterconnects, **(b)** finished IC package

These bonding techniques would be used by semiconductor device manufacturers where assemblers might operate bonding machines. Electronic equipment manufacturers might use the microinterconnect bonding technique to bond IC's to printed-circuit boards as shown in Figure 12-21. The tape of BTP's (Bonding Technique Procedures) is fed through the bonder which chops them out and bonds them to the printed-circuit board. You operate this type of bonder as an electronics assembler.

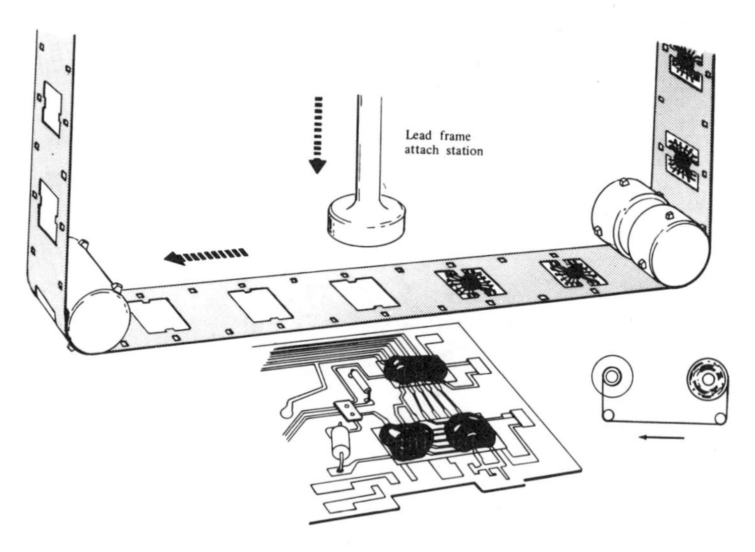

Figure 12-21 Bonding finished IC's and other components to a printed-circuit board

When microinterconnect bonding is used to bond strips to integrated circuits, pads of aluminum or pads with raised bumps must be added to the IC during manufacturing (Figure 12-22). Microinterconnect bonding without these pads results in broken edges on the IC as shown in Figure 12-22(a).

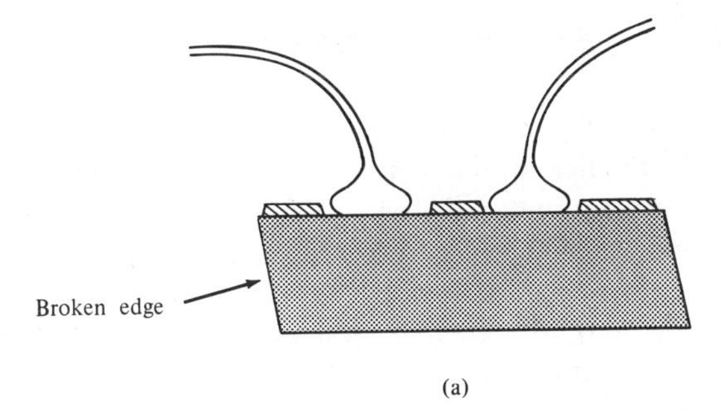

(a)

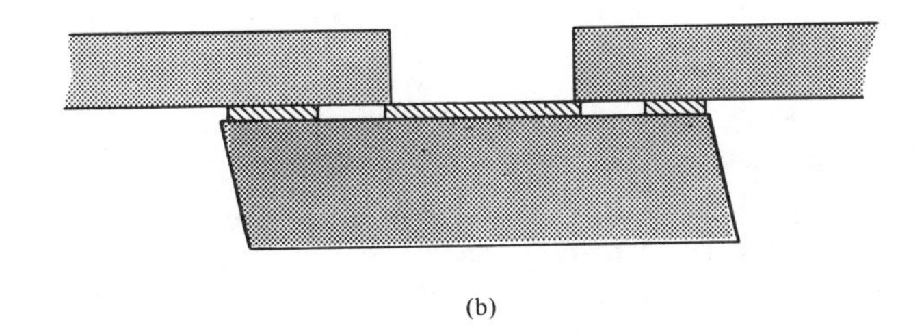

(b)

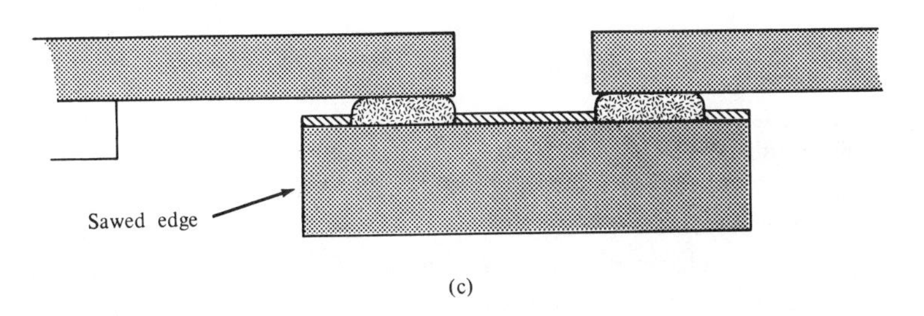

(c)

Figure 12-22 Microinterconnect bonding to IC's: **(a)** without padding, **(b)** with aluminum pads, **(c)** with bumped pads (preferred)

KT's KEY TERMS TO REMEMBER

Adhesive A gluing or bonding material.

Grams per Square Centimeter In the metric system of measurement: the force in grams per centimeter of area.

Noncorrosive Does not weaken or corrode; for example, materials that do not rust or oxidize.

Pounds per Square Inch In the English system of measurement: the force in pounds per inch of area.

Ribbon A small, flat connecting wire.

Thermoplastic A plastic that softens when heated and hardens when cooled. While it is soft, it can be reformed.

TEST

1. Explain the word dwell as it refers to welding.
2. What is the purpose of the electrodes on a welder?
3. What is the purpose of the ribbon in a welding operation?
4. What does the word corrosive mean?
5. What are some of the advantages of hot-melt resins over electrical welding?
6. List the drawbacks to using adhesives.
7. List seven problems that can occur when making a weld.
8. List the general rules to follow when welding.

PROJECT: SPOT WELDING

Materials

1. Spot welder.
2. Several electronic components with small-diameter leads or a 12-in. pin of #20 solid copper wire.
3. A 6-in. lead extension ribbon.

Procedure

Weld a short piece of lead extension ribbon to each lead of the components.

13: Drawing and Blueprint Reading

13.0 INTRODUCTION

Making readable drawings and the reading of schematics, diagrams and blueprints is a requirement for advanced assembler work. These types of drawings will be discussed in this chapter.

13.1 PICTORIAL DRAWINGS

Several forms of representational drawing are used in electronics fabrication and assembly procedures. The simplest type of drawing is a pictorial drawing or diagram, also called a layout diagram, which shows how component parts are assembled in a chassis (Figure 13-1). Various components and terminals may be identified on the drawing by letters or numbers for convenient reference. Pictorial drawings are commonly done in perspective although plan and side views may be used in situations where no confusion will result. Pictorial drawings are often presented as exploded illustrations (Figure 13-2), which show the order in which parts are assembled and their correct relations, when this method will clarify assembly procedures.

Pictorial wiring diagrams (Figure 13-3) are an elaborated variety of the pictorial or layout drawing that include the electrical connections that are made to the component parts following assembly. Wire routes, lead dress, and color codes are presented so pictorial wiring diagrams are an essential guide for the assembler until he or she has learned exactly how to place and connect each lead.

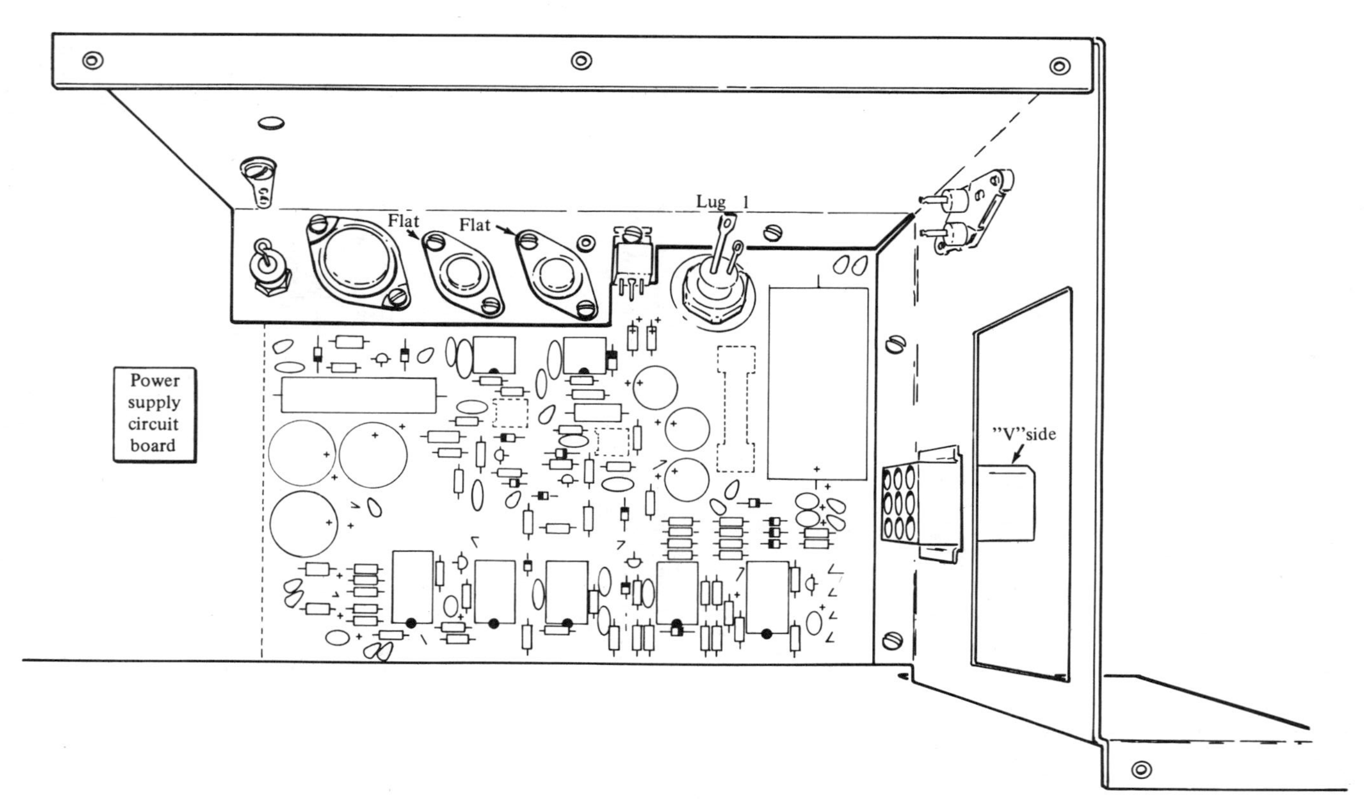

Figure 13-1 A pictorial or layout drawing ***(Courtesy of Heath Company)***

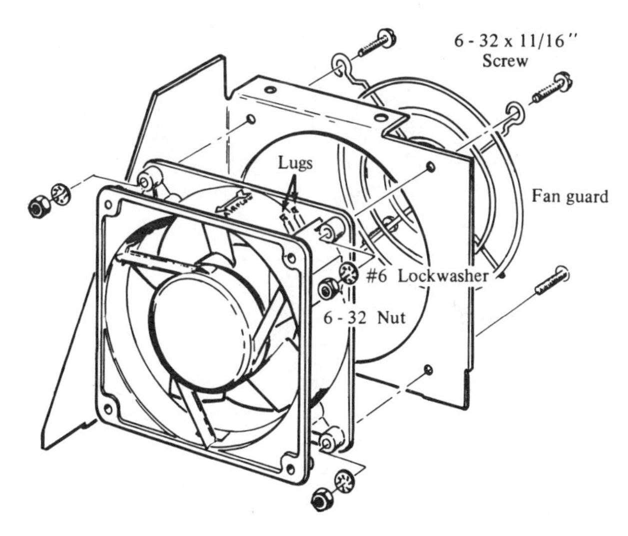

Figure 13-2 An exploded illustration *(Courtesy of Heath Company)*

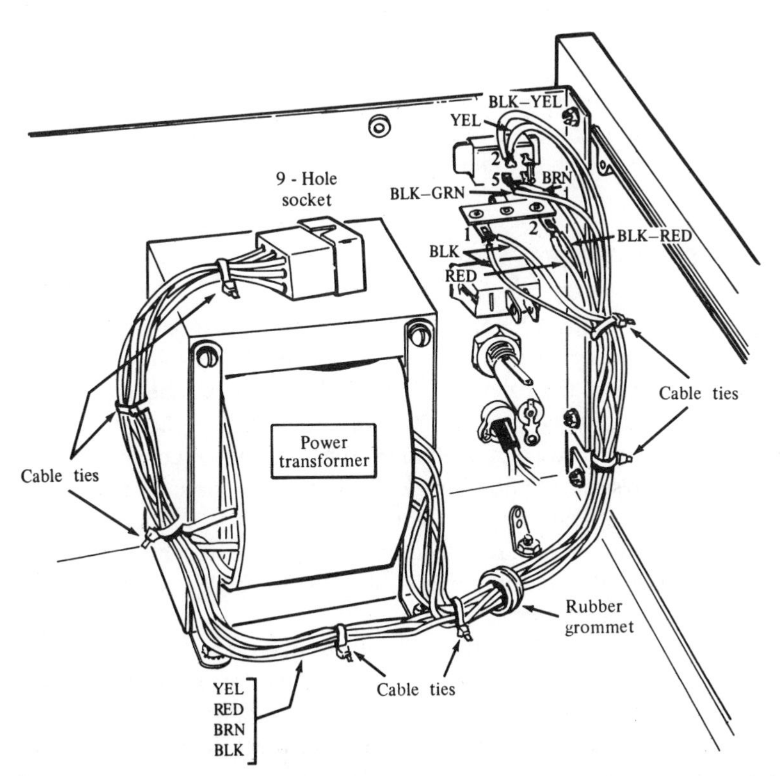

Figure 13-3 A pictorial wiring diagram *(Courtesy of Heath Company)*

The foregoing types of drawings are basically physical renditions in that they show the forms and relationships of hardware and associated items. No conventional symbolization or technical shorthand is employed in this type of drawing. Although pictorial drawings are basic in electronics fabrication and assembly procedures, they have certain limitations and are usually supplemented by other types of drawings and diagrams.

13.2 BLOCK DIAGRAMS

Block diagrams show functional rather than physical relationships. For example, Figure 13-4 shows a block diagram for a television receiver. Each block corresponds to a circuit section, each circuit section has a specific subfunction to perform in the equipment. If a block is drawn in dotted lines, it indicates that that particular subfunction is not utilized in all production models.

Electrical connections are also presented in a block diagram, but in a formalized manner. Single lines with arrowheads denote the progress of electrical signals so the block diagram also acts as a signal flowchart. Block diagrams are not generally used by assemblers but are basic reference diagrams for technicians in final test and troubleshooting stations. Test technicians often utilize block diagrams with reference values such as shown in Figure 13-5.

13.3 SCHEMATIC DIAGRAMS

Schematic diagrams (Figure 13-6) are wiring diagrams that employ standard symbols, a form of technical shorthand that allows for a more rapid preparation than with a pictorial diagram. A schematic diagram may be characterized as an electrical road map although it is less geometric than a conventional map. That is, no attempt is made to show the physical relationships of component parts; the concern is solely with electrical relations from the viewpoint of interconnections. In many cases, a schematic diagram is supplemented by notations regarding electrical and component values, color codes, socket-pin numbers, and switching functions.

Combination schematic and block diagrams also are used in Figure 13-7. A particular section of a television receiver (the intercarrier sound system) is shown in partial schematic detail with the remainder of the receiver sections rendered in block form. Note that integrated circuits are also commonly rendered in block form with numbered

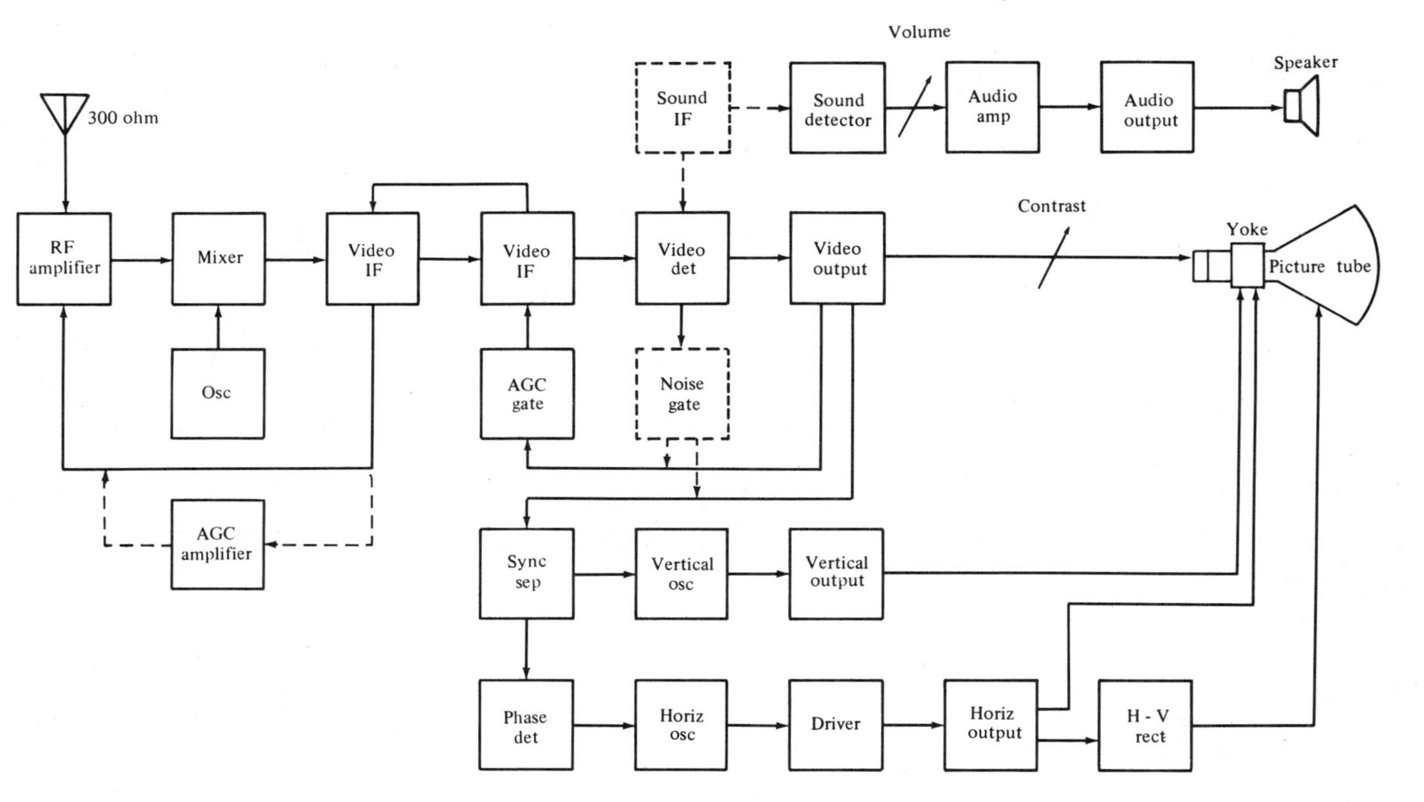

Figure 13-4 Block diagram of a typical transistor television receiver ***(Courtesy of Howard Sams & Company)***

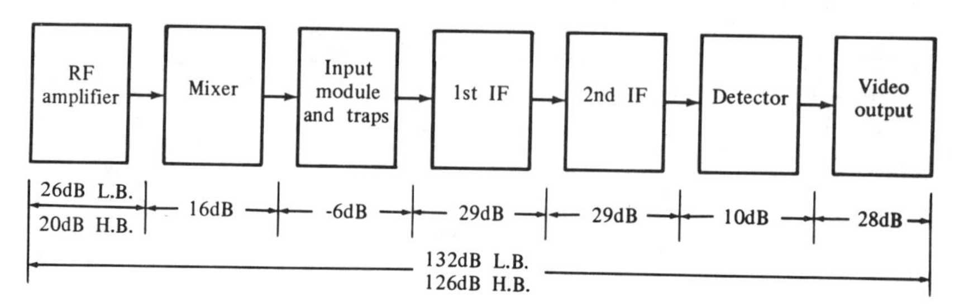

Figure 13-5 A stage-gain block diagram

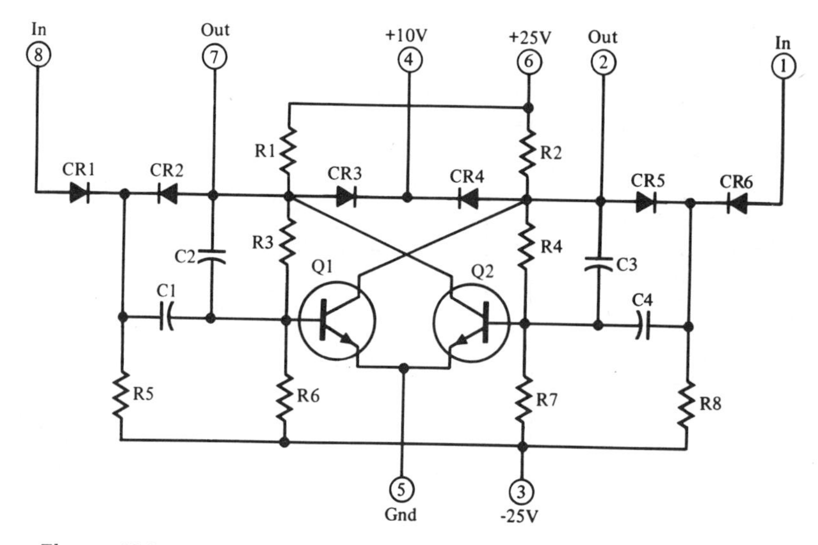

Figure 13-6 A schematic diagram *(Courtesy of Heath Company)*

terminals since these are the data of primary concern to most technicians. An integrated circuit is customarily symbolized by a triangle; thus, since most IC's operate as amplifiers, a triangle is also the standard block representation for an amplifier.

Skeletonized schematic diagrams are sometimes used instead of block diagrams to depict functions, particularly when a single circuit section is involved. For example, Figure 13-8 shows skeletonized representations of transistor amplifier configurations with typical performance characteristics. In the skeletonized schematic, supply voltages and incidental components, such as coupling capacitors, are omitted. The remaining schematic configuration is a definitive presentation of the

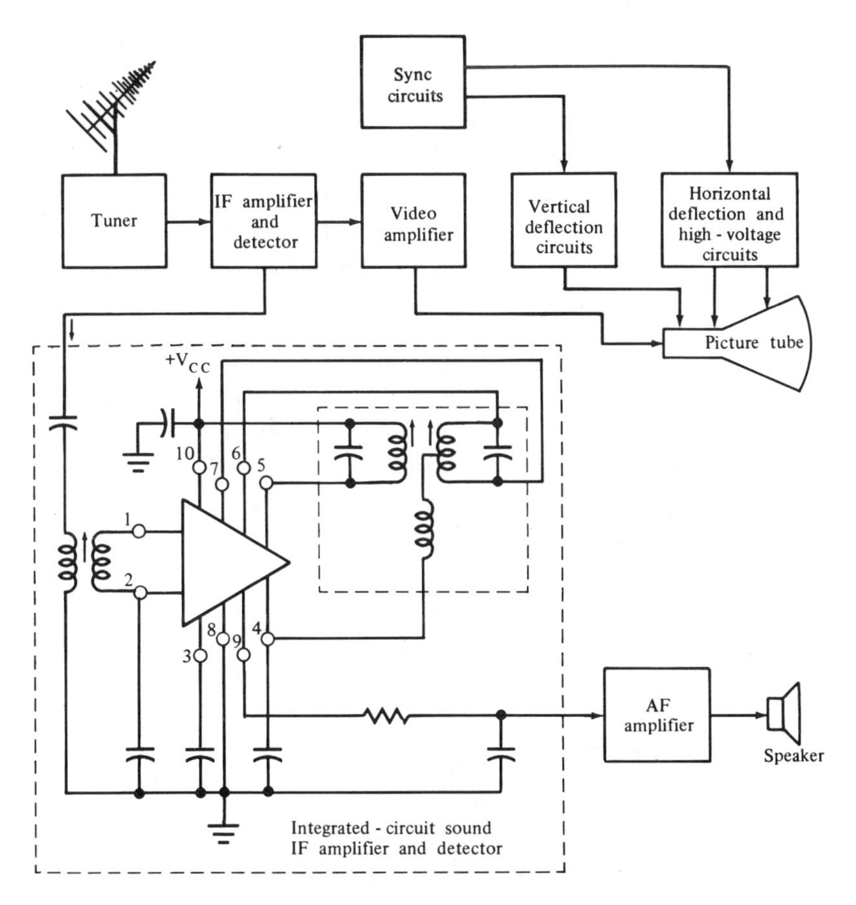

Figure 13-7 An intercarrier sound system that uses an integrated circuit

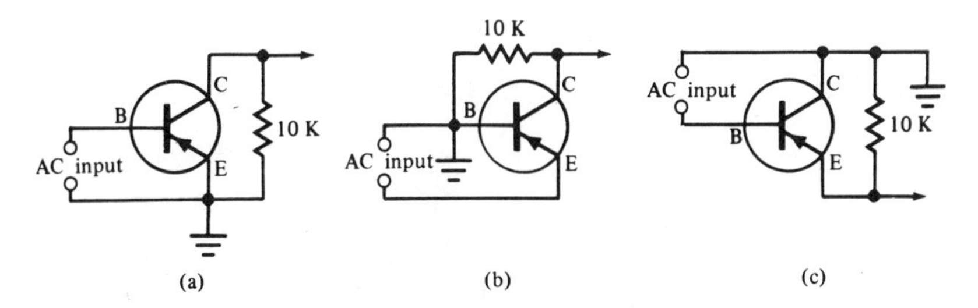

Figure 13-8 Comparison of typical characteristics of a junction transistor in the three basic amplifier configurations

basic electrical design. Definitive diagrams are useful to technicians charged with trouble-shooting malfunctioning equipment.

13.4 UHF "CIRCUIT DIAGRAMS"

Since ultrahigh-frequency (UHF) equipment does not employ circuitry in the usual sense of the term, but operates on the basis of field energy channeled by lines and waveguides, a UHF "circuit diagram" necessarily entails boundary-surface outlines and physical dimensions, as exemplified in Figure 13-9. Lumped components, such as coils and capacitors, are not employed; instead, line or waveguide sections serve analogous functions. Variable tuning is provided by adjustable plungers.

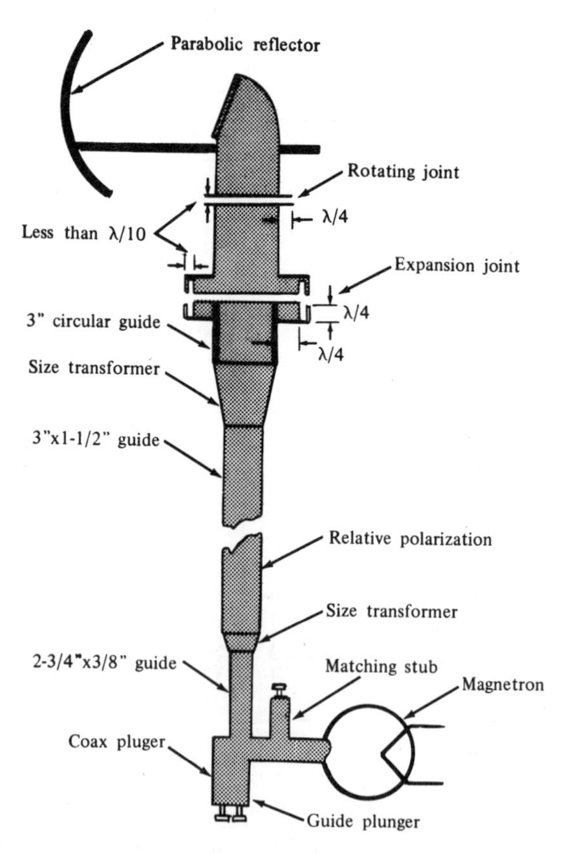

Figure 13-9 UHF "circuit diagram"

13.5 INTERCONNECTION DIAGRAMS

A system often consists of separate units of equipment with specified interconnections. An interconnection diagram shows the wiring configuration for an entire electrical or electronic system. For example, Figure 13-10 depicts the interconnection diagram for a pair of telephone subsets through a common battery exchange. The basic difference between an interconnection diagram and a conventional schematic diagram is that the latter is restricted to a single unit of equipment, whereas an interconnection diagram includes two or more separate units of equipment.

13.6 PROCEDURAL DRAWINGS

Procedural drawings (Figure 13-11) are used to illustrate assembly and/or fabrication techniques. They are employed chiefly in factory manuals to guide apprentice technicians. Another type of procedural drawing is an elaborated form of pictorial wiring diagram in which the wiring involved in a particular step of the procedure is drawn with a particular weight of line (Figure 13-12). The wires that were previously connected are drawn with light lines, and the wires to be connected in the following step are drawn with heavy lines so they stand out prominently. This type of drawing is useful for comparatively complex projects where there is a possibility of confusion among the various steps of the wiring procedure. A similar elaboration is sometimes employed in block diagrams, as exemplified in Figure 13-13. In this drawing, the timing function of the system has been emphasized.

13.7 BLUEPRINTS

Although there is no sharp dividing line between blueprints, line drawings, pictorial drawings, and schematic diagrams, there are two basic distinguishing features. First, a blueprint is rendered as white lines on a blue background. A blueprint may employ either paper or cloth—the blue background results from a photographic emulsion of potassium ferricyanide and ferric salt. Second, a blueprint usually carries dimensional data and is prepared in a standard form observed by professional draftsmen. A blueprint sets forth lines, sections, symbols, dimensions, notes, and titles. All blueprints are examples of mechanical drawings; that is, renditions made with instruments such as compasses, rules, and dividers.

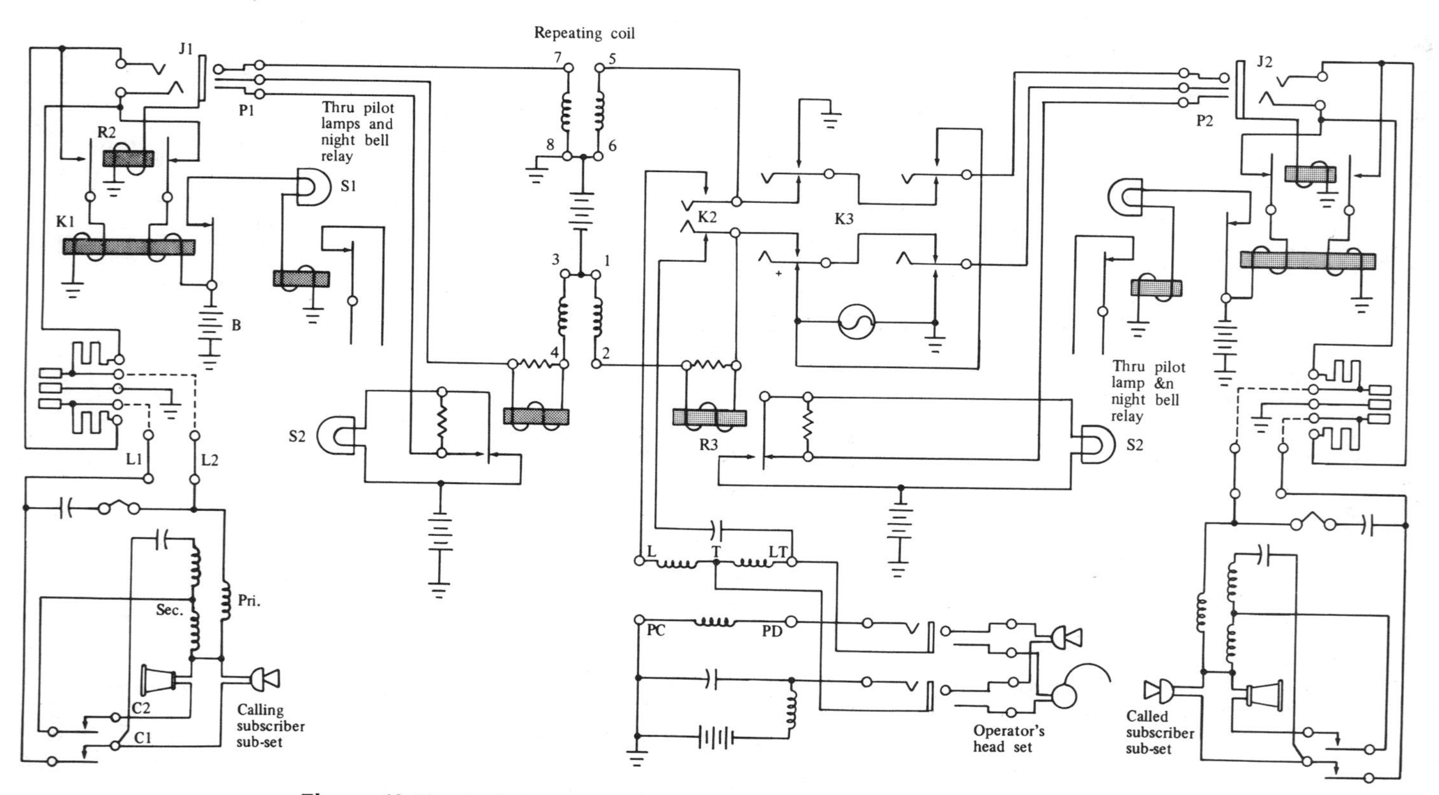

Figure 13-10 An interconnection diagram for a pair of telephone subsets

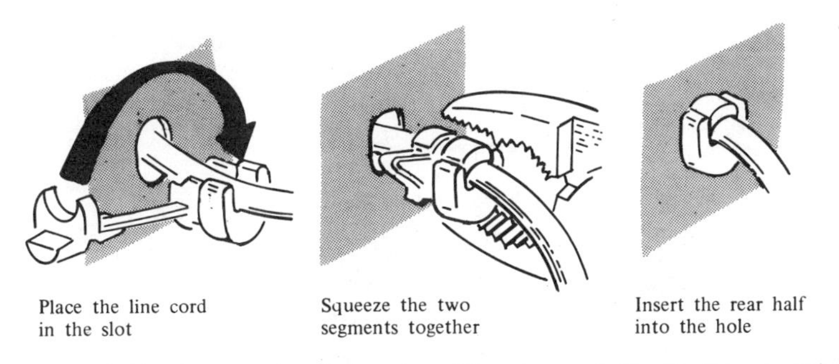

Place the line cord in the slot

Squeeze the two segments together

Insert the rear half into the hole

Figure 13-11 A procedural drawing *(Courtesy of Heath Company)*

The negative for a blueprint, called a tracing, is made by placing a sheet of translucent tracing paper or cloth over the drawing. All the information on the drawing is then copied onto the tracing with black, waterproof ink or a special black pencil. Some drawings are made directly on the tracing sheet in pencil and are then prepared for blueprinting by going over the information with ink or special pencil. The blueprint itself is made by placing the tracing over a sheet of blueprint paper and exposing it to a bright light. The blueprint is developed after exposure by washing in water.

There are many assembly drawings that are similar to blueprints in every way except that the background color is not blue. Black and white prints display black lines on a white background. Ammonia prints, or ozalids, have black, maroon, purple, or blue lines on a white background. Van Dykes have white lines on a dark brown background. Photostats are made as negatives with white lines on a dark black background when only one print is needed. A positive photostat, displaying dark lines on a white background, is made from a negative one when a number of copies are needed.

Let us note the types of blueprints that we will be concerned with. With reference to Figure 13-14, assembling any product begins with a detail blueprint. In this example, the detail blueprint shows an aileron rib (an airplane part). Next, we will find a subassembly blueprint that shows how the aileron rib joins the other parts of the aileron assembly. Next is a unit assembly blueprint. This shows how the aileron joins the other parts of the wing assembly. The final assembly blueprint shows the entire wing assembly in relation to the completed airplane. Plan views are basically maps showing the position, location, and function of the various parts. When a technician needs to make a certain part, he is given a detail print. It may show one large part or several small parts with specifications of size, shape, material, and finishing methods.

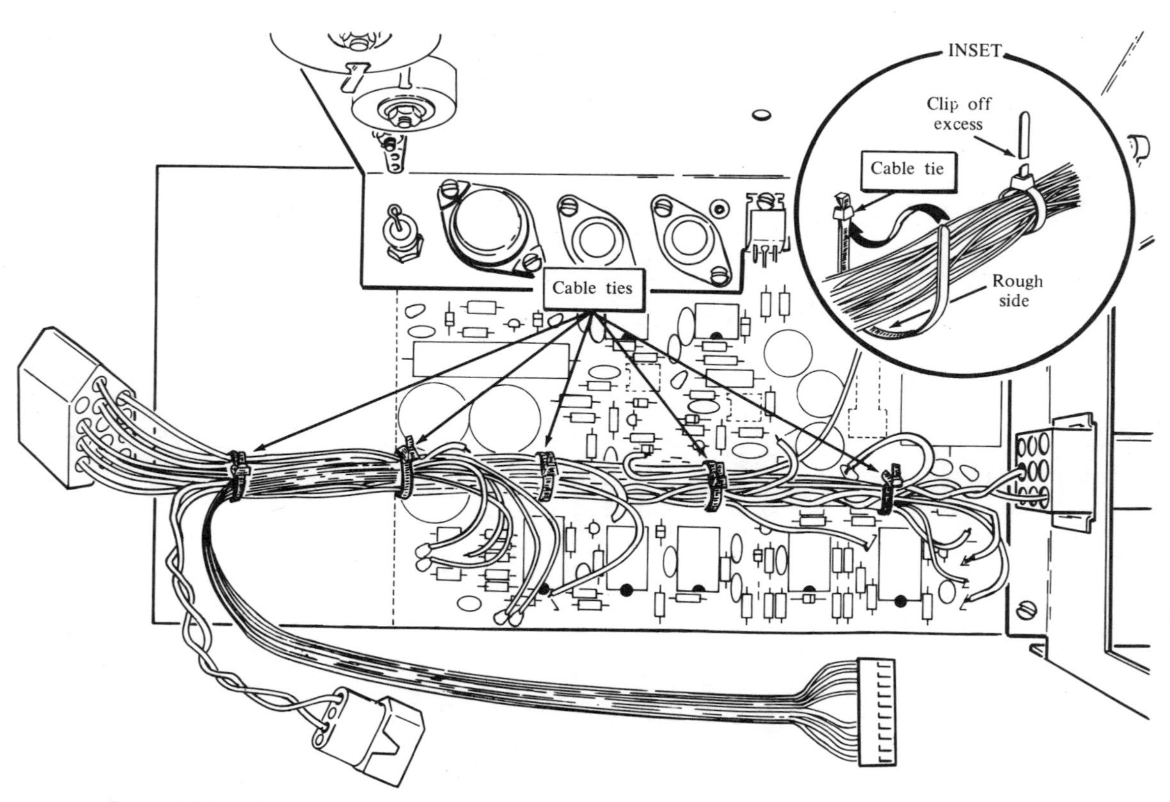

Figure 13-12 A procedural pictorial wiring diagram *(Courtesy of Heath Company)*

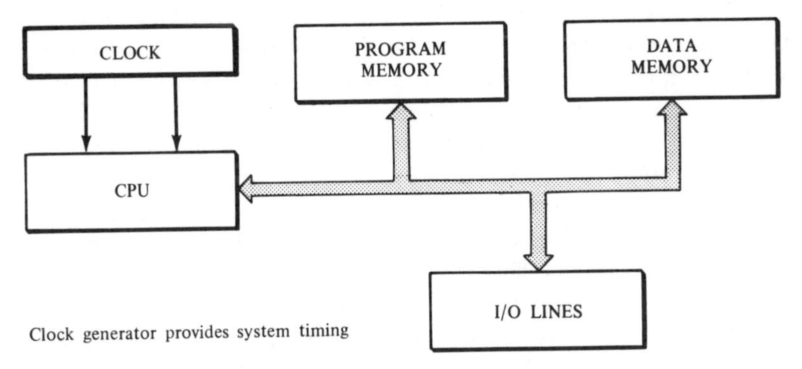

Figure 13-13 Block diagram of rotary spark gap system emphasizing the timing function

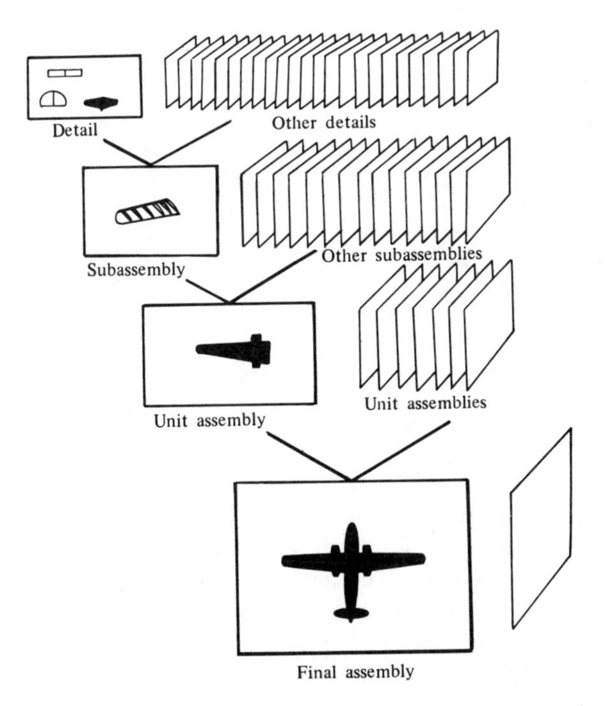

Figure 13-14 Every part, every assembly has its descriptive print

Blueprints should be handled with care. No markings should be made on a print without authorization. A blueprint should not be smudged with grease or placed on a wet surface or left lying in strong sunlight. No attempt should be made to measure sizes or dimensions on a blueprint with a rule; the blueprint will be dimensioned by the draftsmen whenever deemed necessary by the engineering staff. A blueprint is not necessarily drawn precisely to scale, and the paper or cloth base tends to stretch or shrink with changes in humidity and temperature. All blueprints should be filed in their proper place when not in use.

There is a standard method of folding blueprints to ensure that the identifying marks always appear in the same place—preferably at the top when the prints are in vertical filing order. Most of the prints used by technicians will have been previously folded in the correct manner. If a fresh print is to be folded, a number of steps should be followed. All blueprints should be folded neatly in such a manner that the duplicate drawing number appears on the outside in the upper righthand corner of the print. The fold size should be approximately 8½ × 11 in.

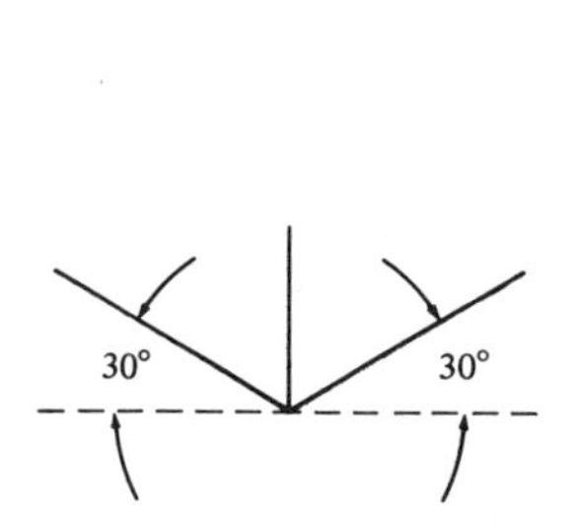

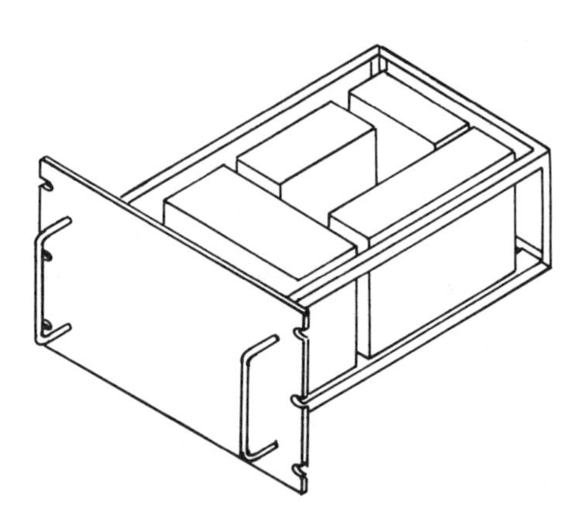

Figure 13-15 Isometric view

Folding blueprints to standard size requires a folding board. This board, fabricated from sheet metal, plastic, or plywood, should be 8⅜ × 10⅞ in. with rounded corners.

A creasing device consisting of a smooth block of wood, metal, or plastic, or a glass paperweight, should be used to press the folds into light creases. Tight, sharp creases will tend to break the sheet and damage it.

On the following sizes of blueprints, straight folds, or no folds, are utilized, as indicated:

1. 8½ × 11 in.—desired size, no fold required
2. 11 × 17 in.—fold once to 11 × 8½ in.
3. 22 × 34 in.—first fold to 34 × 11 in. next to 11 × 17 in., then to 11 × 8½ in.
4. 11 × 34 in.—first fold to 11 × 17 in., then to 11 × 8½ in.

An isometric drawing (Figure 13-15) can be compared in various ways with a photograph and a perspective drawing. Basically, an isometric drawing is simplified and formalized type of perspective drawing. In isometrics, lines that are actually parallel on the object are rendered parallel on the drawing, although the simulated perspective rendition is distorted. All lines representing horizontal and vertical lines on the object have true lengths. Vertical lines are shown in a vertical position, but horizontal lines in the object are rendered at an angle of 30° with the horizontal, in an isometric. Vertical lines and lines representing horizontal lines in the object are called isometric lines. In Figure 13-15, all horizontal and vertical lines have true lengths.

Blueprints may also provide auxiliary projections. With reference to Figure 13-16, note the inclined surface shown in the front view, the right side view and the top view, the inclined surface appears foreshortened; its true size is not represented in the rendition. Therefore, the drafter adds an auxiliary view made with respect to a perpendicular presentation of the inclined surface. Curved surfaces do not necessarily appear as curved surfaces in an orthographic projection. For example, Figure 13-17, shows a two-view orthographic drawing of a solid cone. A third view, which would ordinarily appear above the bottom view is not necessary because it would be a duplicate of the side view and would provide no additional information. We know the sides of a cone are curved, but this curvature is not obvious. Therefore, an interpretation is involved in reading the blueprint.

13.8 SUMMARY

Technicians who are concerned with electronic fabrication and assembly procedures must become familiar with various types of drawings and blueprints. The most basic type of technical drawing is a pictorial, or layout drawing, presented in perspective. A pictorial drawing may be shown as an exploded illustration which clearly indicates the spatial order and relations of individual parts. A pictorial wiring diagram is a pictorial drawing with connecting wires included. Block diagrams show

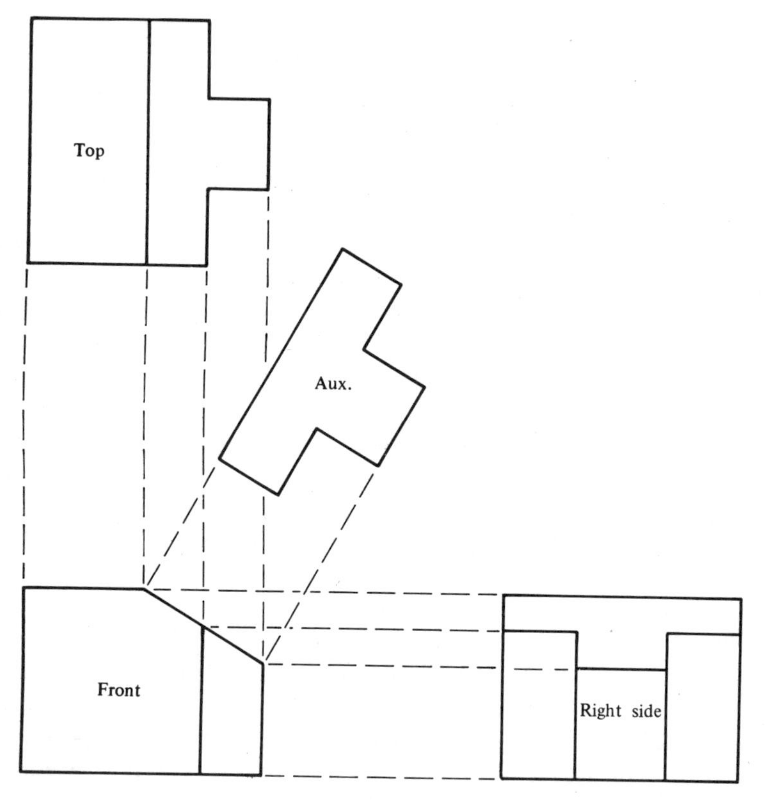

Figure 13-16 Orthographic views with an auxiliary view

the chief circuit sections with interconnecting arrows that provide a signal flowchart. Block diagrams may have electrical values indicated at key test points, such as the input or output terminals of various sections.

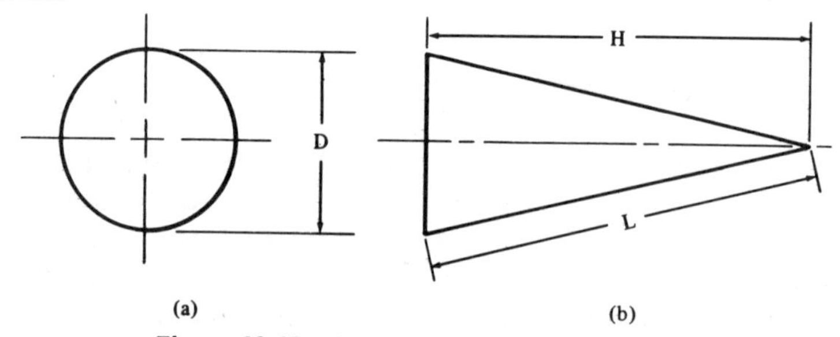

Figure 13-17 Orthographic views of a cone

Schematic diagrams show electrical connections and employ standard graphical symbols. A schematic might be termed an electrical road map in technical shorthand form. Combination block and schematic diagrams set forth electrical design characteristics, and omit incidental electrical connections, supply voltages, and circuit details. UHF circuit diagrams are a specialized form of pictorial drawing that shows boundary surfaces and their relations.

An interconnection diagram shows how individual units of equipment are connected together into a system. Procedural drawings illustrate essential points in fabrication or assembly procedures. They include procedural pictorials and modified block diagrams.

A set of blueprints includes a print for every part and assembly in an equipment unit or system. Standard procedures are observed in folding blueprints. Various projections, other than perspective, are utilized in blueprint drawings. Isometric projection, which is a simplified and modified form of perspective, is common. Orthographic projection employs three perpendicular views of an object. Since an inclined surface cannot be rendered in true proportions by orthographic projection, it is common practice to include an auxiliary view or projection of an inclined surface. Curves do not necessarily appear as such in orthographic projections; interpretation is required in such cases when reading a blueprint.

LABORATORY EXERCISES

The following laboratory exercises provide practical experience for the student concerning the reading of drawings and blueprints.

EXERCISE 1: PREPARING A BASIC PICTORIAL DRAWING

Objectives

1. To become familiar with basic pictorial drawings
2. To learn how to make simple technical sketches
3. To recognize basic assembly requirements

Materials and Equipment

Sketch pad, ruler, drawing pencil, coordinate paper, small power supply with components mounted

Procedure

1. Evaluate the requirements for a pictorial drawing and decide whether more than one view will be desirable.
2. Make a perspective pictorial drawing or drawings of the equipment.
3. Identify each component by a letter and append a list to show the technical nomenclature of the components.

Review Questions

1. When is a single pictorial drawing sufficient?
2. Describe a situation in which two or more pictorial drawings would be desirable.
3. Why is it helpful to make a pictorial drawing in perspective?

EXERCISE 2: PREPARING A BASIC PICTORIAL WIRING DIAGRAM

Objectives

1. To learn to observe wiring routes and connection details.
2. To become familiar with pictorial wiring diagrams.
3. To learn the essentials of an informative drawing.

Materials and Equipment

Sketch pad, ruler, drawing pencil, coordinate paper, small pre-wired power supply.

Procedure

1. Evaluate the requirements for a pictorial drawing and decide whether more than one view will be desirable.
2. Make a perspective pictorial drawing or drawings of the equipment.
3. Note any color coding or special features of the wiring plan.

Review Questions

1. Why might several progressive pictorial wiring diagrams be desirable for an elaborate electronic unit?
2. Suggest several purposes that can be served by color-coded wires.
3. Why might a procedural drawing be desirable to supplement a pictorial wiring diagram?

EXERCISE 3: PREPARING A SCHEMATIC DIAGRAM

Objectives

1. To become familiar with basic schematic diagrams.
2. To learn how to draw simple schematic diagrams.
3. To learn to recognize the essentials of schematic presentation.

Materials and Equipment

Sketch pad, ruler, compass, drawing pencil, prewired power supply.

Procedure

1. Study the wiring of the unit and make a rough sketch of the wiring plan.
2. Rearrange your rough sketch to present the information in standard schematic form.
3. Make a neat schematic diagram from your rearranged rough sketch.

Review Questions

1. Why is it impractical to proceed directly from observing an electronic unit to a finished schematic diagram?
2. When a choice can be made, would you select a layout of graphical symbols that minimizes the number of conductor cross-overs? Why or why not?
3. Is it desirable to show the input terminals at the left-hand, right-hand, top, or bottom of a schematic diagram? Where is it desirable to place the output terminals? What are the reasons for your answers?

PROJECT: BLUEPRINT AND SCHEMATIC READING

Materials

1. Schematic diagram
2. Chassis drawing
3. Drafting layout of a chassis
4. Sheet of 8 × 12 in. drawing paper

Procedure

1. Verbally identify each component on the schematic diagram furnished by your instructor.
2. Verbally discuss each measurement on the chassis drawing furnished by your instructor.
3. Correctly make a flat drawing of the chassis in Figure 13, showing all dimensions.
4. Make a schematic drawing from a pictorial drawing furnished by your instructor.

Index